OPTICAL AND MICROWAVE ENGINEERING

Deepthi Paramasivam

Assistant Professor

Department of Electronics and Communication Engineering

Government College of Engineering Salem, Tamilnadu, India.

SHANLAX #SINCE1990

SHANLAX PUBLICATIONS

Title:	**OPTICAL AND MICROWAVE ENGINEERING**
Author's Name:	**Deepthi Paramasivam**
Published by:	**Shanlax Publications, Vasantha Nagar, Madurai - 625003, Tamil Nadu, India**
Publisher's Address:	**61, 66 T.P.K. Main Road, Vasantha Nagar, Madurai - 625003, Tamil Nadu, India**
Printer's Details:	**Shanlax Press, 66 T.P.K. Main Road, Vasantha Nagar, Madurai - 625003, Tamil Nadu, India**
Edition Details (I,II,III):	**I**
ISBN:	**978-93-93737-02-1**
Month & Year:	**January, 2022**
Pages:	**218**
Price:	**390/-**

ACKNOWLEDGEMENT

First and foremost, praise and thanks to God, the Almighty, our Parents and our Family Members, and our well-wishers for their showers of blessings throughout our book work and to complete the publication successfully.

I have to thank my awesome husband, Ajai. He was as important to this book getting done as I was. Thank you so much, dear.

Thanks to everyone on the publishing team
Last but not least we are thankful to all our students who allow us to learn more and more.
We hope readers will enjoy this book and if any suggestions, always feel welcome.

Deepthi Paramasivam

Semester VII 18EC701	Optical and Microwave Engineering	L T P C 3 0 0 3

UNIT I MICROWAVE COMPONENTS 9

Hybrid Circuits - Waveguide Tees -Magic Tees (Hybrid Tees) – Hybrid Rings (Rat-Race Circuits) – Waveguide Corners - Bends and Twists - Directional Couplers – Two-Hole Directional Couplers – Review oflow frequency parameters: Z,Y and ABCD Parameters – Introduction to S parameters – S Matrix of a Directional Coupler – Hybrid Couplers - Circulators and Isolators.

UNIT II SOLID STATE MICROWAVE DEVICES 9

Introduction- Gunn_Effect Diodes – GaAs Diode – Ridely-Watkins – Hilsun (RWH) Theory –Modes of Operation - Microwave Generation and Amplification - Avalanche transit - Time devices Introduction - Read Diode – IMPATT Diodes - TRAPATT Diodes -BARITT Diodes – Parametric Devices.

UNIT III MICROWAVE TUBES 9

Klystrons - Two cavity Klystron Amplifiers - Reflex Klystrons – Velocity Modulation - Power Output and Efficiency - Electronic Admittance - Helix Traveling - Wave Tubes (TWTs)

- Slow-Wave structures – Amplification Process - Convection Current - Axial Electric Field – Wave Modes - Gain Consideration - Magnetron Oscillators - Cylindrical Magnetron - Coaxial Magnetron.

UNIT IV SIGNAL DEGRADATION IN OPTICAL FIBERS 9

Attenuation - Absorption losses – Scattering losses - Bending Losses – Core and Cladding losses - Signal Distortion in Fibers - Intermodal delay - Intramodal dispersion – Factors contributing to dispersion – GroupDelay - Material Dispersion - Wave guide Dispersion - Basics of semiconductor physics LED Structures - Light source materials – Quantum efficiency and LED power - LASER diodes

UNIT V FIBER OPTICAL RECEIVERS AND DIGITAL TRANSMISSION SYSTEM 9

Physical principles of photodiodes – PIN photo diode - Avalanche photo diodes - Photodetector noise - SNR- Detector response time - Double Hetero structure photodiodes - structure for InGaAS APDs -Temperature effect on avalanche gain – Fundamental receiver operation – Digital signal transmission – Error sources – Frontend amplifier - Digital receiver performance - Receiver sensitivity - Optical Amplifiers Types - Erbium Doped fiber amplifier.

INDEX

UNIT - I

MICROWAVE COMPONENTS

Hybrid Circuits - Waveguide Tees - Magic Tees (Hybrid Tees) - Hybrid Rings (Rat-Race Circuits) - Waveguide Corners - Bends and Twists - Directional Couplers - Two-Hole Directional Couplers - Review of low frequency parameters: Z,Y and ABCD Parameters - Introduction to S parameters - S Matrix of a Directional Coupler - Hybrid Couplers - Circulators and Isolators.

1.1 Hybrid Circuits

An electronic circuit composed of different types of integrated circuits and discrete components, mounted on a ceramic base. Used in military and communications applications, it is especially suited for building custom analog circuits including A/D and D/A converters, amplifiers and modulators. The hybrid microcircuit evolved into the multichip module (MCM) and, later, the multichip package (MCP).

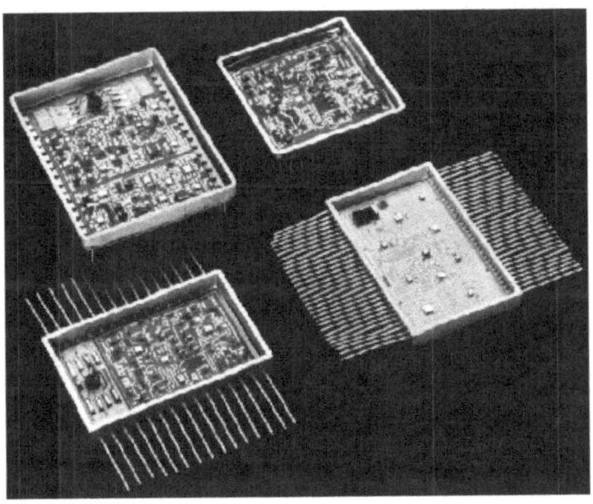

Figure: Hybrid Microcircuits

The picture shows a variety of hybrid circuits. The tiny, square white spots are the actual chips.

1.2 Waveguide Tees

A Waveguide Tee is a 3-port device that can be used to either divide or combine power in a waveguide system. It is formed when three waveguides tubes are connected in the form of the English alphabet 'T'. This is where its name is derived from.

Figure: E – Plane Tee & H – Plane Tee

There are two main types of Waveguide Tees:

1. E-Plane Waveguide Tee

A waveguide tee is a 3 port device that is similar to a power divider. When the axis of the side arm is parallel to the Electric Field (E) of the collinear, then the tee is called a E-Plane Tee Junction. The outputs we get in this type of tee are 180° out of phase with each other, irrespective of from which port the input is fed.

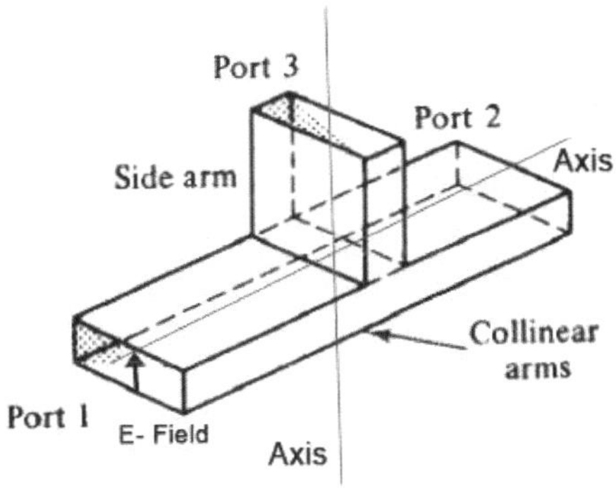

Figure: E-Plane Waveguide Tee

2. H-Plane Waveguide Tee

When the axis of the side arm of the waveguide tee is parallel to the flow of the Magnetic Field (H) from port 1 and is perpendicular to the flow of the Electric Field (E), then the tee is called a H-Plane Waveguide Tee.

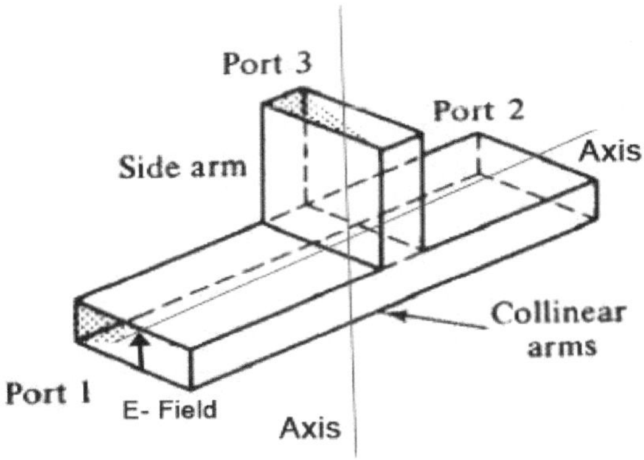

Figure: H-Plane Waveguide Tee

An H-Plane Waveguide Tees can be thought of as a two way in-phase power divider/combiner i.e it is additive in nature. When two input signals are fed to port 1 & 2, the output at port 3 is in phase and additive and when the input signal is fed to port 3, the signal is split in to two equal parts that are in-phase at port 1 & 2.

1.3 Magic Tees (Hybrid Tees)

A Magic Tee or Hybrid Tee is a 4 port waveguide tee that is a combination of an E-Plane and H-Plane Waveguide Tee.

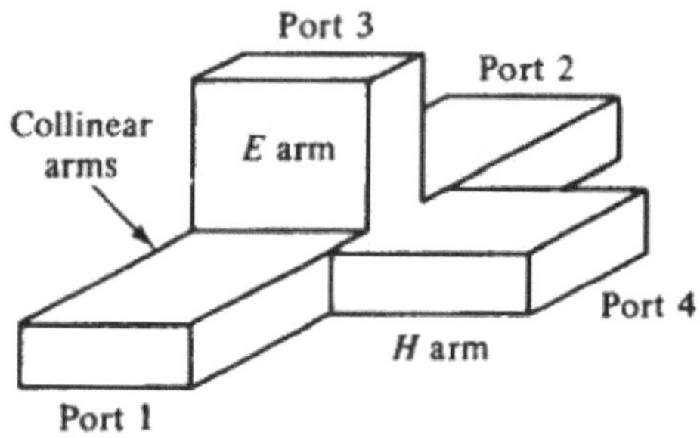

Magic Tee (Hybrid Tees)

Figure: Magic Tees (Hybrid Tees)

A magic tee has four ports

- Port 1 - Co-linear
- Port 2 - Co-linear
- Port 3 - Difference Port
- Port 4 - Sum Port

Operation of a Magic Tee

Case 1: When two signals of equal magnitude are fed from port 1 and 2, we get a zero at port 3 and the sum of the two signals at port 4.

Case 2: When a signal is fed through port 4, it gets divided equally between port 1 & 2 and both the outputs are in phase. No output comes from port 3.

Case 3: When a signal is fed through port 3, we get an output of equal magnitude but opposite phase at port 1 & 2 (the signals are 180 degrees out of phase). Output at port 4 is zero.

A magic tee is ideally lossless. But the biggest disadvantage of magic tee is that reflections arise within it due to impedance mismatches, which causes some level of power loss. These reflections can be minimized by optimizing matching.

1.4 Hybrid Rings (Rat-Race Circuits)

This microwave device is used when there is a need to combine two signals with no phase difference and to avoid the signals with a path difference.

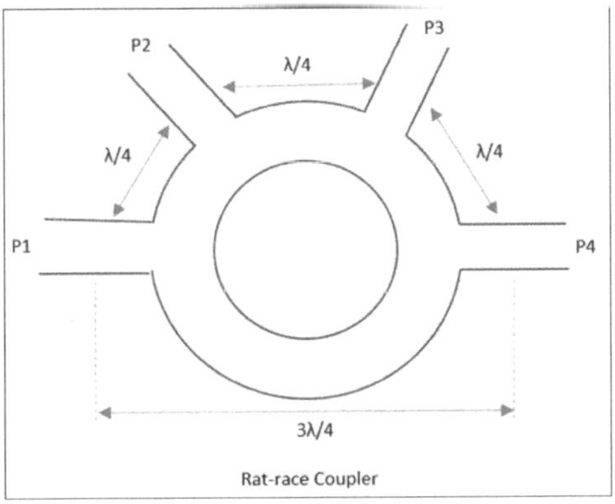

Figure: Hybrid ring

A normal three-port Tee junction is taken and a fourth port is added to it, to make it a ratrace junction. All of these ports are connected in angular ring forms at equal intervals using series or parallel junctions.

The mean circumference of total race is 1.5λ and each of the four ports are separated by a distance of λ/4. The following figure shows the image of a Rat-race junction.

Let us consider a few cases to understand the operation of a Rat-race junction.

Case 1: If the input power is applied at port 1, it gets equally split into two ports, but in clockwise direction for port 2 and anti-clockwise direction for port 4. Port 3 has absolutely no output.

The reason being, at ports 2 and 4, the powers combine in phase, whereas at port 3, cancellation occurs due to $\lambda/2$ path difference.

Case 2: If the input power is applied at port 3, the power gets equally divided between port 2 and port 4. But there will be no output at port 1.

Case 3: If two unequal signals are applied at port 1 itself, then the output will be proportional to the sum of the two input signals, which is divided between port 2 and 4. Now at port 3, the differential output appears.

The Scattering Matrix for Rat-race junction is represented as

$$[S] = \begin{bmatrix} 0 & S_{12} & 0 & S_{14} \\ S_{21} & 0 & S_{23} & 0 \\ 0 & S_{32} & 0 & S_{34} \\ S_{41} & 0 & S_{43} & 0 \end{bmatrix}$$

Applications

Rat-race junction is used for combining two signals and dividing a signal into two halves.

1.5 Waveguide Corners, Bends and Twists

The waveguide corner, bend, and twist are shown in Figure. These waveguide components are normally used to change the direction of the guide through an arbitrary angle.

In order to minimize reflections from the discontinuities, it is desirable to have the mean length L between continuities equal to an odd number of quarter-wavelengths. That is

$$L = (2n + 1)\frac{\lambda_g}{4}$$

where, n = 0, 1, 2, 3, ... , and Ag is the wavelength in the waveguide. If the mean length L is an odd number of quarter wavelengths, the reflected waves from both ends of the waveguide section are completely canceled. For the waveguide bend, the minimum radius of curvature for a small reflection is given by :

$$R = 1.5b \qquad \text{for an } E \text{ bend}$$

$$R = 1.5a \qquad \text{for an } H \text{ bend}$$

where, a and b are the dimensions of the waveguide bend as illustrated

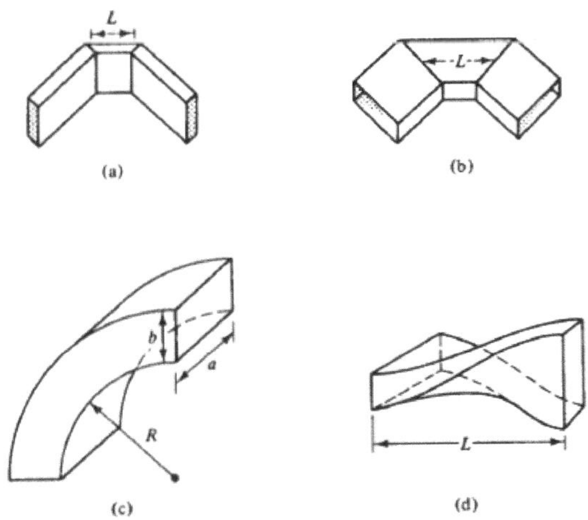

Figure: Waveguide corner, bend, and twist. (a) E-plane corner. (b) H-plane corner. (c) Bend. (d) Continuous twist

1.6 Types of waveguide bend

1.6 Types of waveguide bend

There are several ways in which waveguide bends can be accomplished. They may be used according to the applications and the requirements.

- Waveguide E bend
- Waveguide H bend
- Waveguide sharp E bend
- Waveguide sharp H bend

Each type of bend is achieved in a way that enables the signal to propagate correctly and with the minimum of disruption to the fields and hence to the overall signal.

Ideally the waveguide should be bent very gradually, but this is normally not viable and therefore specific waveguide bends are used.

Most proprietary waveguide bends are common angles - 90° waveguide bends are the most common by far.

Waveguide E bend

This form of waveguide bend is called an E bend because it distorts or changes the electric field to enable the waveguide to be bent in the required direction.

To prevent reflections this waveguide bend must have a radius greater than two wavelengths.

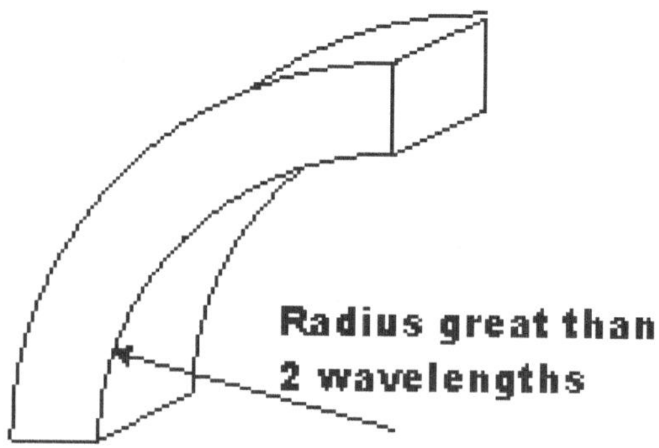

Waveguide H bend

This form of waveguide bend is very similar to the E bend, except that it distorts the H or magnetic field. It creates the bend around the thinner side of the waveguide.

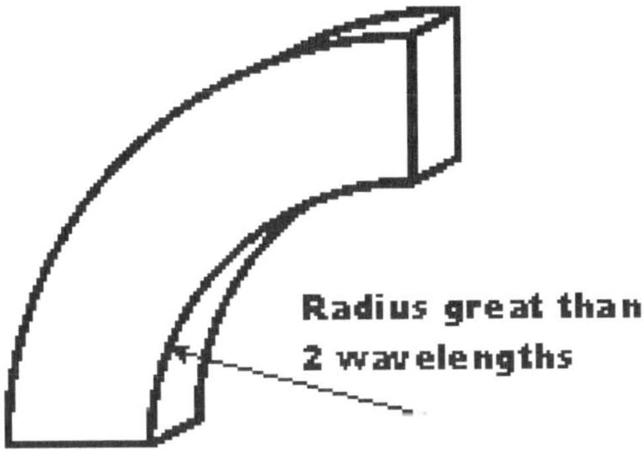

As with the E bend, this form of waveguide bend must also have a radius greater than 2 wavelengths to prevent undue reflections and disturbance of the field.

Waveguide sharp E bend

In some circumstances a much shorter or sharper bend may be required. This can be accomplished in a slightly different manner. The techniques is to use a 45° bend in the waveguide. Effectively the signal is reflected, and using a 45° surface the reflections occur in such a way that the fields are left undisturbed, although the phase is inverted and in some applications this may need accounting for or correcting.

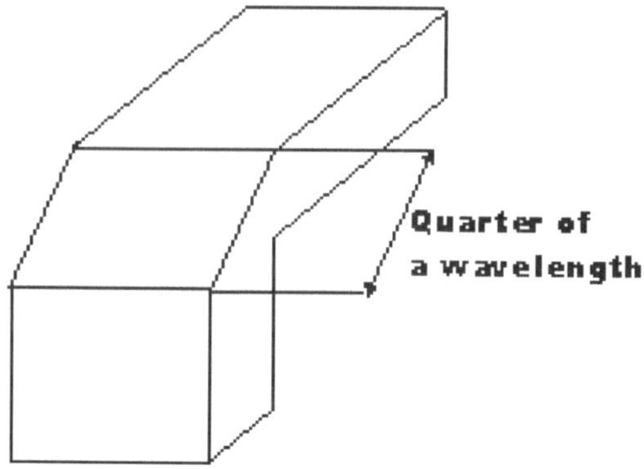

Waveguide sharp H bend

This for of waveguide bend is the same as the sharp E bend, except that the waveguide bend affects the H field rather than the E field.

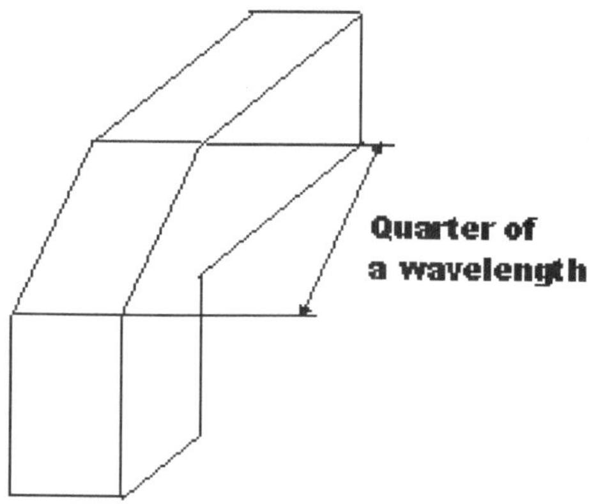

Waveguide Twists

There are also instances where the waveguide may require twisting. This too, can be accomplished. A gradual twist in the waveguide is used to turn the polarisation of the waveguide and hence the waveform.

In order to prevent undue distortion on the waveform a 90° twist should be undertaken over a distance greater than two wavelengths of the frequency in use. If a complete inversion is required, e.g. for phasing requirements, the overall inversion or 180° twist should be undertaken over a four wavelength distance.

Waveguide bends and waveguide twists are very useful items to have when building a waveguide system. Using waveguide E bends and waveguide H bends and their srap bend counterparts allows the waveguide to be turned through the required angle to meet the mechanical constraints of the overall waveguide system. Waveguide twists are also useful in many applications to ensure the polarisation is correct.

1.7 Directional Couplers

A Directional coupler is a device that samples a small amount of Microwave power for measurement purposes. The power measurements include incident power, reflected power, VSWR values, etc.

Directional Coupler is a 4-port waveguide junction consisting of a primary main waveguide and a secondary auxiliary waveguide. The following figure shows the image of a directional coupler.

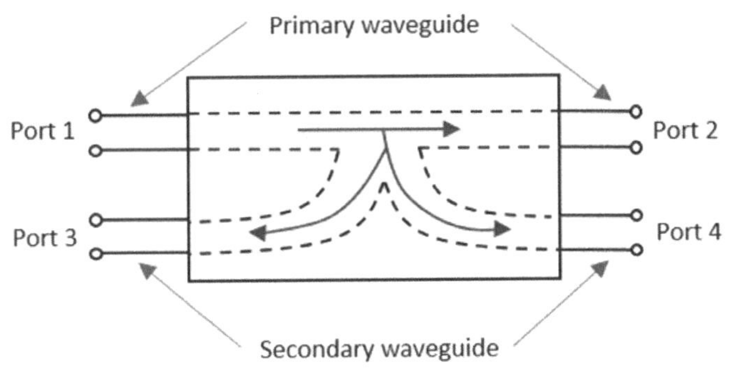

Directional Coupler

Directional coupler is used to couple the Microwave power which may be unidirectional or bi-directional.

Properties of Directional Couplers

The properties of an ideal directional coupler are as follows.

- All the terminations are matched to the ports.
- When the power travels from Port 1 to Port 2, some portion of it gets coupled to Port 4 but not to Port 3.
- As it is also a bi-directional coupler, when the power travels from Port 2 to Port 1, some portion of it gets coupled to Port 3 but not to Port 4.
- If the power is incident through Port 3, a portion of it is coupled to Port 2, but not to Port 1.
- If the power is incident through Port 4, a portion of it is coupled to Port 1, but not to Port 2.
- Port 1 and 3 are decoupled as are Port 2 and Port 4.

Ideally, the output of Port 3 should be zero. However, practically, a small amount of power called back power is observed at Port 3. The following figure indicates the power flow in a directional coupler.

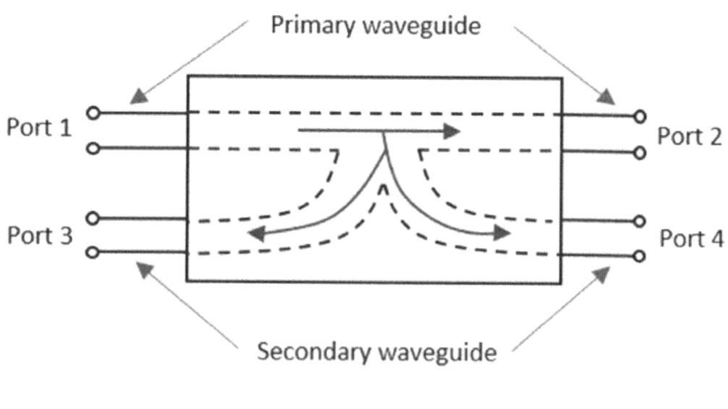

Directional Coupler

Where,

- Pi = Incident power at Port 1
- Pr = Received power at Port 2
- Pf = Forward coupled power at Port 4
- Pb = Back power at Port 3

Following are the parameters used to define the performance of a directional coupler.

Coupling Factor C

The Coupling factor of a directional coupler is the ratio of incident power to the forward power, measured in dB.

$$C = 10\,log_{10}\frac{P_i}{P_f}\,dB$$

Directivity D

The Directivity of a directional coupler is the ratio of forward power to the back power, measured in dB.

$$D = 10\,log_{10}\frac{P_f}{P_b}\,dB$$

Isolation

It defines the directive properties of a directional coupler. It is the ratio of incident power to the back power, measured in dB.

$$I = 10\,log_{10}\frac{P_i}{P_b}\,dB$$

Isolation in dB = Coupling factor + Directivity

1.8 Two-Hole Directional Coupler

This is a directional coupler with same main and auxiliary waveguides, but with two small holes that are common between them. These holes are λg/4 distance apart where λg is the guide wavelength. The following figure shows the image of a two-hole directional coupler.

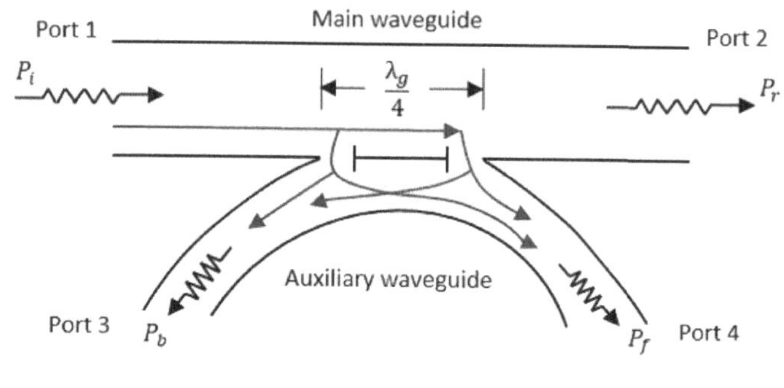

Two-hole directional coupler

A two-hole directional coupler is designed to meet the ideal requirement of directional coupler, which is to avoid back power. Some of the power while travelling between Port 1 and Port 2, escapes through the holes 1 and 2.

The magnitude of the power depends upon the dimensions of the holes. This leakage power at both the holes are in phase at hole 2, adding up the power contributing to the forward power Pf. However, it is out of phase at hole 1, cancelling each other and preventing the back power to occur.

Hence, the directivity of a directional coupler improves.

Waveguide Joints

As a waveguide system cannot be built in a single piece always, sometimes it is necessary to join different waveguides.

This joining must be carefully done to prevent problems such as – Reflection effects, creation of standing waves, and increasing the attenuation, etc.

The waveguide joints besides avoiding irregularities, should also take care of E and H field patterns by not affecting them. There are many types of waveguide joints such as bolted flange, flange joint, choke joint, etc.

1.9 Review of Low Frequency Parameters: Z,Y and ABCD Parameters

The ABCD-parameters are known variously as chain, cascade, or transmission line parameters. There are a number of definitions given for ABCD parameters, the most common

$$\begin{bmatrix} V_1 \\ I_1 \end{bmatrix} = \begin{bmatrix} A & B \\ C & D \end{bmatrix} \begin{bmatrix} V_2 \\ -I_2 \end{bmatrix}$$

For reciprocal networks AD – BC =1 For symmetrical networks A = D . For networks which are reciprocal and lossless, A and D are purely real while B and C are purely imaginary.

This representation is preferred because when the parameters are used to represent a cascade of two-ports, the matrices are written in the same order that a network diagram would be drawn, that is, left to right. However, the examples given below are based on a variant definition

$$\begin{bmatrix} V_2 \\ I'_2 \end{bmatrix} = \begin{bmatrix} A' & B' \\ C' & D' \end{bmatrix} \begin{bmatrix} V_1 \\ I_1 \end{bmatrix}$$

where

$$A' \stackrel{\text{def}}{=} \left.\frac{V_2}{V_1}\right|_{I_1=0} \qquad B' \stackrel{\text{def}}{=} \left.\frac{V_2}{I_1}\right|_{V_1=0}$$

$$C' \stackrel{\text{def}}{=} \left.-\frac{I_2}{V_1}\right|_{I_1=0} \qquad D' \stackrel{\text{def}}{=} \left.-\frac{I_2}{I_1}\right|_{V_1=0}$$

The negative signs in the definitions of parameters C‘ and D‘ arise because I2‘ is defined with the opposite sense to I2 , that is,I2‘ = I2 . The reason for adopting this convention is so that the output current of one cascaded stage is equal to the input current of the next. Consequently, the input voltage/current matrix vector can be directly replaced with the matrix equation of the preceding cascaded stage to form a combined A‘B‘C‘D‘ matrix.

The terminology of representing the parameters as a matrix of elements designated a11 etc as adopted by some authors [10] and the inverse parameters as a matrix of elements designated b11 etc is used here for both brevity and to avoid confusion with circuit elements.

1.10 Introduction to S Parameters

Before we get into the math, let's define a few things you need to know about S-parameters.

The scattering matrix is a mathematical construct that quantifies how RF energy propagates through a multi-port network. The S-matrix is what allows us to accurately describe the properties of incredibly complicated networks as simple "black boxes". For an RF signal incident on one port, some fraction of that signal gets reflected back out of the incident port, some of it enters *into* the incident port and then exits at (or *scatters* to) some or all of the other ports (perhaps being amplified or attenuated). What's left of that incident power disappears as heat or even electromagnetic radiation. The S-matrix for an N-port contains N^2 coefficients (S-parameters), each one representing a possible input-output path.

S-parameters are complex numbers, having real and imaginary parts or magnitude and phase parts, because both the magnitude and phase of the incident signal are changed by the network. Quite often we refer to the magnitude of the signal only, as it is frequently of most interest. Who cares how the signal phase is changed by an amplifier or attenuator? You mostly care about how much gain (or loss) you get. S-parameters are defined for a given frequency and system impedance, and vary as a function of frequency for any non-ideal network.

S-parameters are usually displayed in a matrix format, with the number of rows and columns equal to the number of ports. For the S-parameter S_{ij} the j subscript stands for the port that is excited (the input port), and the "i" subscript is for the output port. Thus S_{11} refers to the ratio of the amplitude of the signal that reflects from port one to the amplitude of the signal incident on port one. Parameters along the diagonal of the S-matrix are referred to as reflection coefficients because they only refer to what happens at a single port, while off-diagonal S-parameters are referred to as transmission coefficients, because they refer to what happens at one port when it is excited by a signal incident at another port. Here are the S-matrices for one, two and three-port networks:

$$(S_{11}) \quad (one-port)$$

$$\begin{pmatrix} S_{11} & S_{12} \\ S_{21} & S_{22} \end{pmatrix} \quad (two-port)$$

$$\begin{pmatrix} S_{11} & S_{12} & S_{13} \\ S_{21} & S_{22} & S_{23} \\ S_{31} & S_{32} & S_{33} \end{pmatrix} \quad (three-port)$$

Etc.

Note that each S-parameter is a complex number, so if actual data were presented in matrix format, a magnitude and phase angle would be presented for each S_{ij}.

The input and output reflection coefficients of networks (such as S_{11} and S_{22}) can be plotted on the Smith chart. Transmission coefficients (S_{21} and S_{12}) are usually not plotted on the Smith chart.

1.11 Definition of S-parameters

S-parameters describe the response of an N-port network to signal(s) incident to any or all of the ports. The first number in the subscript refers to the responding port, while the second number refers to the incident port. Thus S_{21} means the response at port 2 due to a signal at port 1. The most common "N-port" networks in microwaves are one-port and two-port networks. Three-port network S-parameters are easy to model with software such as Agilent ADS, but three-port S-parameter measurements are extremely difficult to perform with accuracy. Measured multi-port S-parameters are typically available from vendors for amplifiers and other devices but, as always, make sure you check your answers for reasonableness.

Let's examine a two-port network. The signal at a port, say port 1, can be thought of as the superposition to two waves traveling in opposite directions. By convention each port is shown as two nodes so as to give a name and value to these opposite direction waves. The variable a_i represents a wave incident to port i and the variable b_j represent a wave reflected from port j. Don't get all hung up on how two signals can occur at the same node! The magnitude of the a_i and b_j variables can be thought of as voltage-like variables, normalized using a specified reference impedance. This is very convenient since the square of these magnitudes are then equal to the power level of the waves. Remember, S-parameters don't mean much unless you know the value of the reference impedance (it's frequently called Z_0).

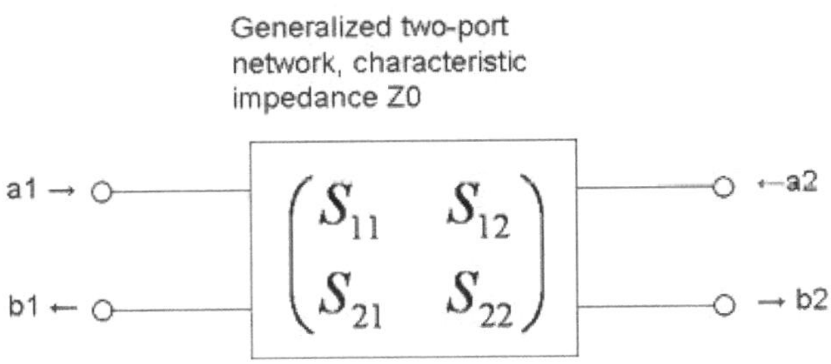

If we assume that each port is terminated in the reference impedance Z_0, we can define the four S-parameters of the 2-port as:

$$S_{11} = {b1}/{a1}$$

$$S_{12} = {b1}/{a2}$$

$$S_{21} = {b2}/{a1}$$

$$S_{22} = {b2}/{a2}$$

The above equations for S_{11} and S_{21} are derived from network analysis or measurements by setting the value of the incident signal $a_2=0$ and solving for the above S-parameter ratios as a function of a_1. Similarly, S_{12} and S_{22} are derived by setting the value of $a_1=0$ and solving for the other ratios.

See how the subscript neatly follows the parameters in the ratio ($S_{11}=b_1/a_1$, etc...)? Here's the matrix algebraic representation of 2-port S-parameters:

$$\begin{pmatrix} b_1 \\ b_2 \end{pmatrix} = \begin{pmatrix} S_{11} & S_{12} \\ S_{21} & S_{22} \end{pmatrix} \times \begin{pmatrix} a_1 \\ a_2 \end{pmatrix}$$

If we want to measure S_{11}, we inject a signal at port one and measure its reflected signal. In this case, no signal is injected into port 2, so a2=0; for almost all laboratory S-parameter measurements, we only inject one signal at a time. If we want to measure S_{21}, we inject a signal at port 1, and measure the resulting signal power exiting port 2. For S_{12} we inject a signal into port 2, and measure the signal power leaving port 1, and for S_{22} we inject a signal at port 2 and measure its reflected signal.

Did we mention that all the a and b measurements are complex numbers? It isn't always necessary to keep track of the angle of the S-parameters, but complex S-parameters are a much more powerful tool than magnitude-only S-parameters, and the math is simple enough either way. By the way, these complex numbers are sometimes called vectors, hence the term vector network analyzers (VNA).

S-parameter magnitudes are presented in one of two ways, linear magnitude or logarithmic based decibels (dB). Because S-parameters are complex voltage ratios, the formula for decibels in this case is

$$S_{ij}(dB)=20 * \log[S_{ij}(magnitude)]$$

Remember that power ratios are expressed as 10 * log(whatever). Voltage ratios are 20 * log(whatever), because power is proportional to voltage squared.

The angle or phase of a complex S-parameter is almost always presented in degrees (but, of course, radians are possible).

Types of S-parameters

When we are talking about networks that can be described with S-parameters, we are usually talking about single-frequency networks. Receivers and mixers aren't referred to as having S-parameters, although you can certainly measure the reflection coefficients at each port and refer to these parameters as S-parameters. The trouble comes when you wish to describe the frequency-conversion properties, this is not possible using S-parameters.

- **Small signal S-parameters** are what we are talking about 99% of the time. By small signal, we mean that the signals have only linear effects on the network, small enough so that gain compression or other non-linear effects do not take place. For passive networks, small-signal is all you have to worry about, because they act linearly at any power level (at least until you blow them up).
- **Large signal S-parameters** are more complicated. In this case, the S-matrix will vary depending upon the input signal strength. Measuring and modeling large signal S-parameters will not be described on this page (perhaps we will get into that someday)

- **Mixed-mode S-parameters** refer to a special case of analyzing balanced circuits. We're not going to get into that either!
- **Pulsed S-parameters** are measured on power devices so that an accurate representation is captured before the device heats up. This is a tricky measurement, and not something we're gonna tackle yet.

S Matrix of a Directional Coupler

Directional Coupler is a 4 port device which has primary and secondary waveguides. The primary wave guide is from port 1 to port 2 and secondary waveguide is from port 3 to port 4.

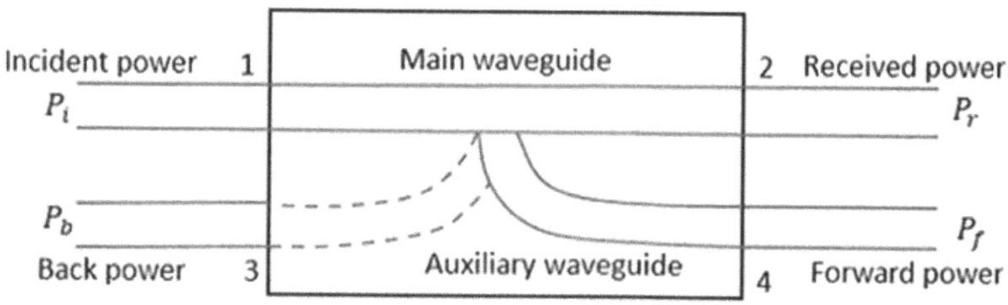

- Directional coupler is used to couple microwave power, which is unidirectional in most of the cases. The properties of a directional coupler are,
  1. All the ports are matchead.
  2. When the power moves from port 1 to port 2, little portion of it would gets coupled to port 4 and not to port 3.
  3. When the power moves from port 2 to port 1, little portion of it would gets coupled to port 3 and not to port 4.
  4. The coupling factor of a directional coupler is the ratio of incident power to forward power.

Coupling factor = 10 log(Pi/Pf)

- Directivity of the directional coupler is the ratio of forward power to back power.

Directivity = 10 log(Pf/Pb)

- Isolation of a directional coupler is the ratio of incident power to back power.

I = 10 log(Pi/Pb)

- Isolation = Coupling factor + Directivity.
- Two hole directional coupler is same as conventional directional coupler, but with two holes in common between primary and secondary waveguides.

The spacing between these two holes is given by,

L = (2n+1)λg/4

Where, n = an integer

λg = wavelength

- A fraction of energy entering into port 1 passes through holes and is radiated into port 2. The forward waves in port 4 are in the same phase and are added.
- The backward waves in port 3 are out of phase and are cancelled.

The general S matrix of a directional coupler is,

$$S = \begin{bmatrix} S_{11} & S_{12} & S_{13} & S_{14} \\ S_{21} & S_{22} & S_{23} & S_{24} \\ S_{31} & S_{32} & S_{33} & S_{34} \\ S_{41} & S_{42} & S_{43} & S_{44} \end{bmatrix} \text{-------} (1)$$

- Since all ports in a directional coupler are matched.

$$S_{11} = S_{22} = S_{33} = S_{44} = 0 \text{ ----------------- } (2)$$

- Since there is no coupling between ports 1 & 3 and ports 2 & 4

$$S_{13} = S_{31} = S_{24} = S_{42} = 0 \text{ ------------- } (3)$$

Apply equation (2) & (3) in (1)

$$S = \begin{bmatrix} 0 & S_{12} & 0 & S_{14} \\ S_{12} & 0 & S_{23} & 0 \\ 0 & S_{23} & 0 & S_{34} \\ S_{14} & 0 & S_{34} & 0 \end{bmatrix}$$

By unitary property, [S][S]* = I

$$\begin{bmatrix} 0 & S_{12} & 0 & S_{14} \\ S_{12} & 0 & S_{23} & 0 \\ 0 & S_{23} & 0 & S_{34} \\ S_{14} & 0 & S_{34} & 0 \end{bmatrix} \begin{bmatrix} 0 & S_{12}^* & 0 & S_{14}^* \\ S_{12}^* & 0 & S_{23}^* & 0 \\ 0 & S_{23}^* & 0 & S_{34}^* \\ S_{14}^* & 0 & S_{34}^* & 0 \end{bmatrix} = \begin{bmatrix} 1 & 0 & 0 & 0 \\ 0 & 1 & 0 & 0 \\ 0 & 0 & 1 & 0 \\ 0 & 0 & 0 & 1 \end{bmatrix}$$

R1C1 => $|S_{12}|^2 + |S_{14}|^2 = 1$ ------------ (4)

R2C2 => $|S_{12}|^2 + |S_{23}|^2 = 1$ ------------ (5)

R3C3 => $|S_{23}|^2 + |S_{34}|^2 = 1$ ------------ (6)

R1C3 => $S_{12} S_{23}^* + S_{14} S_{34}^* = 0$ ----------- (7)

Comparing eq (4) and (5)

$|S_{12}|^2 + |S_{14}|^2 = |S_{12}|^2 + |S_{23}|^2$

$S_{14} = S_{23}$ --------------- (8)

Comparing eq (5) and (6)

$|S_{12}|^2 + |S_{23}|^2 = |S_{34}|^2 + |S_{23}|^2$

$S_{12} = S_{34}$ --------------- (9)

Let, S12 be real and positive,

ie, $S_{12} = S_{34} = p$ --------------- (10)

applying equation (10) in (7)

Therefore, $p\, S_{23}^* + S_{14}\, p = 0$

$p\, [S_{23}^* + S_{14}] = 0$

$p\, [S_{23}^* + S_{23}] = 0$

$S_{23}^* + S_{23} = 0$

To satisfy the above condition, S23 should be a complex value.

Let $S_{23} = jq$

Therefore, the S matrix of directional coupler is,

$$S = \begin{bmatrix} 0 & p & 0 & jq \\ p & 0 & jq & 0 \\ 0 & jq & 0 & p \\ jq & 0 & p & 0 \end{bmatrix}$$

1.12 Hybrid Couplers

Hybrid Coupler Basics & Selection

A hybrid coupler is a four port directional coupler that splits an input signal equally between two output ports, with an attenuation of 3 dB. The signals at the outputs have a 90 degree or 180 degree phase difference between them.

90° Hybrid Couplers

A 90° Hybrid Coupler is a 4 port device that is used to equally split an input signal with a 90° phase shift between the ports. To illustrate operation, a signal at the input port (port 1) is divided between the two output ports with half the power flowing to the 0° port and the other half flowing to the 90° port. Any reflection from mismatches at the output port flow to the Isolated port (port 4). This is one of the reasons hybrid couplers are used to split signals in applications where unwanted reflections can damage the circuit. This type of coupler is also known as a Quadrature coupler.

Everything RF has listed 90° Hybrid Couplers from the leading manufacturers - API, MECA, Mini-Circuits, IPP, Werlatone etc. Use the parametric search tool to narrow down on the products - Download Datasheets and Request Quotations. These are directed to the manufacturers or reps in your region who will get back to you with the details. Click Here to start searching.

180° Hybrid Couplers

A 180° Hybrid Coupler is a four port device that is used either to equally split an input signal with a 180° phase shift between the ports or to combine two signals that are 180° apart in phase. 180° Hybrid Couplers usually consist of a center conductor ring with a circumference of 1.5 times the wavelength (6 times the quarter wavelength). Each port is separated by a quarter wavelength (90° apart). This configuration creates a low loss device with low VSWR and excellent phase and amplitude balance. This type of coupler is also know as —rat race coupler‖.

Everything RF has listed 180° Hybrid Couplers from the leading manufacturers - API, Mini-Circuits, PMi, IPP, Werlatone etc. Use the parametric search tool to narrow down on the products - Download Datasheets and Request Quotations. These are directed to the manufacturers or reps in your region who will get back to you with the details. Click Here to start searching.

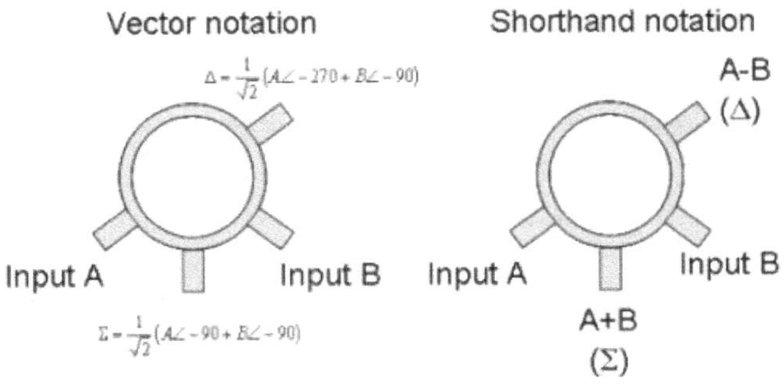

Ratrace used as a combiner

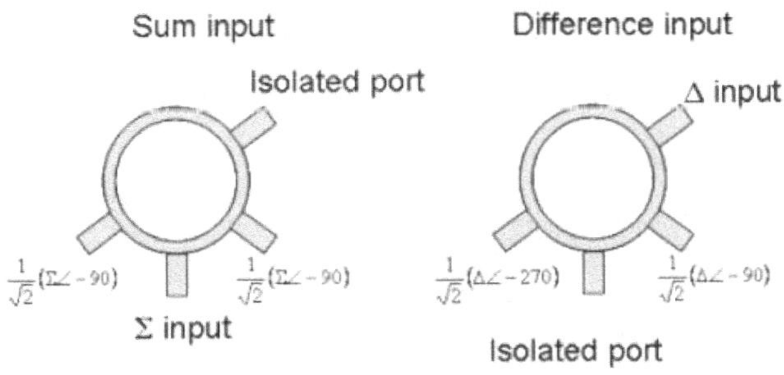

Rat-race used as a divider

Hybrid Coupler Selection

When selecting a hybrid coupler it is important to know the parameters that you need to specify. The key parameters would be Type, Frequency, Insertion Loss, Isolation and Power.

1.13 Circulators and Isolators

RF or microwave circulators or isolators are devices that typically have three and sometimes four ports and they are used in RF system designs that require power from one port to be transferred to another whilst isolating power from the other.

RF circulators are used in many RF applications acting as a duplexer, allowing both transmit and receive functions to occur simultaneously, they are widely used in RF design applications including radar systems and a variety of professional radio communications systems.

RF circulators are receive their name because they transfer power from one port to the next, circulating it from say, entering at port one to output at port two, and entering at port two to exit at port three.

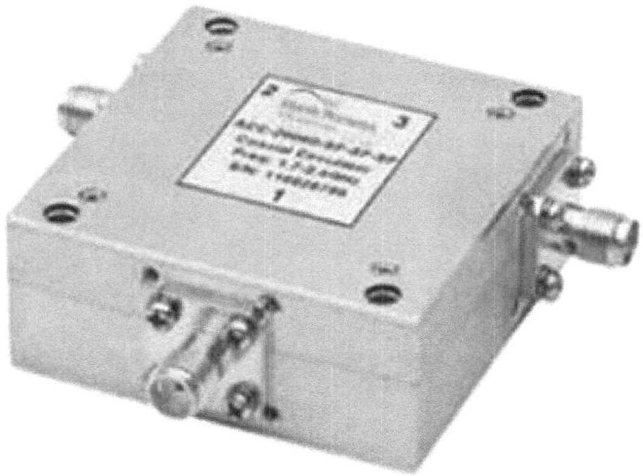

Typical microwave circulator this one is from Atlantic Microwave

Typical RF circulator

How an RF circulator works

The connections to RF circulators are normally called ports, and in addition to this they are normally numbered as 1, 2, 3, etc.

The RF circulator gains its name because it circulates the power entering one port only to the next one. A signal applied to port 1 will be passed to port 2: a signal input to port 2 will pass to port 3, but not back to port 1. An input to port 3 will pass to port 1, but not in reverse to port 2.

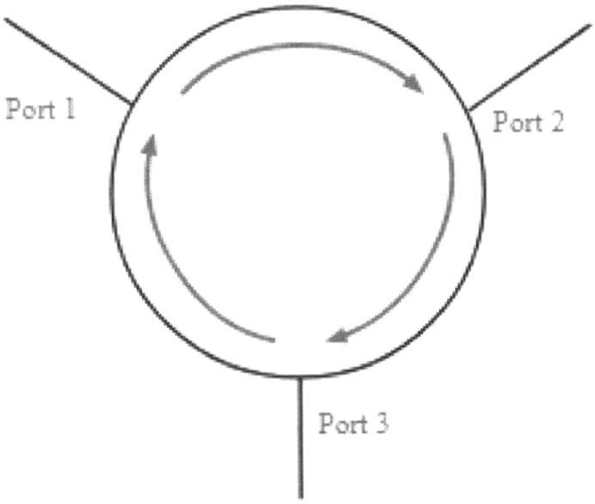

Concept of a basic RF circulator used for microwave RF circuit design

RF Circulator Concept

The ideal circulator will transfer all the power from one port to the required port, and none to any other. However in reality, there is always some attenuation in the transfer path, and some signal always leaks onto the ports that should be isolated. The key RF circuit design challenge for these devices is to ensure the optimum transfer and isolation occurs.

Circulators may use strip-line printed circuit board technology (but normally using very low loss dielectric or PCB materials) and be contained within metal boxes with connectors or other connections to the outside world - some even use surface mount technology. Other circulators may be based within waveguides and these can be used in RF system design applications that incorporate waveguide technology. The type of interface and technology that are need for any given instance will depend upon the RF circuit design for the application. In terms of their operation there most RF circulators are based around the use of ferromagnetic material. There are two main types:

Three Port Circulators

3-port "Y-junction" circulators based on cancellation of waves propagating over two different paths near a magnetised material. Waveguide circulators may be of either type, while more compact devices based on strip-line are of the 3-port type.

Four Port Circulators

4-port waveguide circulators based on Faraday rotation of waves propagating in a magnetised material. Using this technology, they are able to route the RF signals to four ports.

RF Circulator Circuit Symbol

Lie other electronic components, the RF circulator has its own circuit symbol that is used to represent it on electronic circuit diagrams or schematics.

The basic symbol consists of the circle and an arrow showing the direction in which the power is circulated. Normally the ports are showin in order clockwise around the circle: Port 1, Port 2, and finally Port 3.

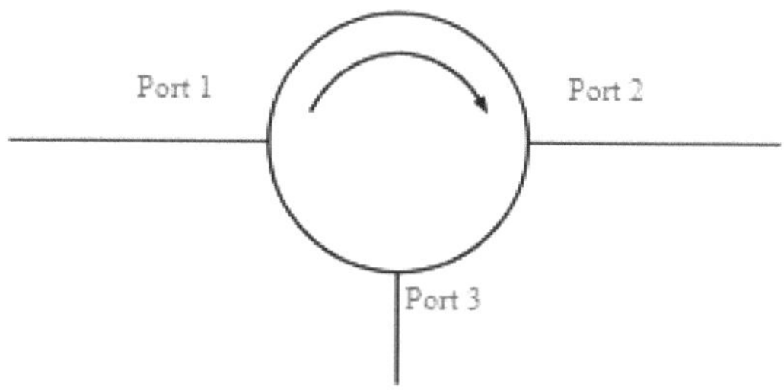

RF circulator circuit symbol

Circuit Symbol for a circulator

It is worth noting that each port, whether a coaxial feeder or waveguide, is shown as a single line rather than a pair of conductors.

3 port Stripline Technology Circulators

One of the most commonly used forms of RF circulator is formed from a Y-shaped section of micro-strip or strip-line transmission line that is on a printed circuit board or other dielectric. The ports are placed 120° apart so that they are equi-spaced around a circle.

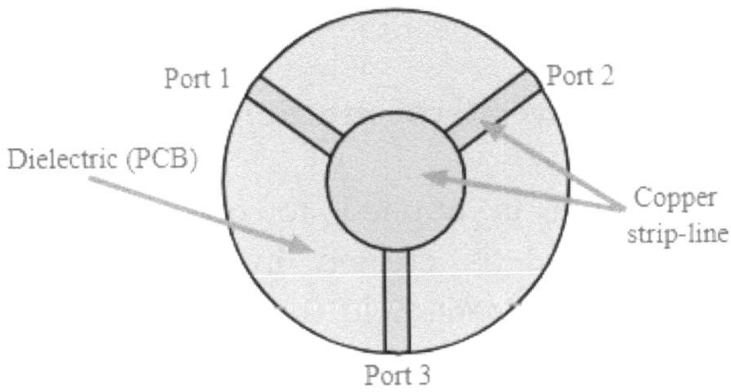

The format of a Y shaped strip-line RF circulator printed circuit board

The printed circuit board assembly is then sandwiched between two laters of ferrite and then outside this, two strong magnets are fixed in place.

The assembly sets up a strong magnetic field axially through the ferrite discs and this focuses the magnetic field, called the bias, around the Y-junction.

Overall construction of an RF circulator showing the cross section with the ferrite discs and magnets

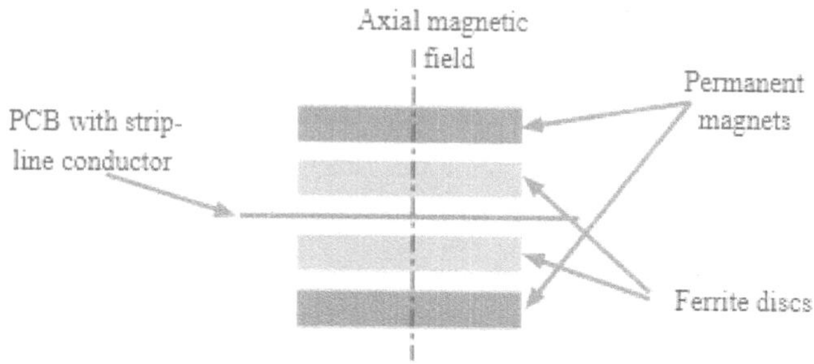

Cross section of overall construction of the RF circulator

When a signal is applied to one of the ports, an electromagnetic field is created in the strip-line, which interacts with the magnetic field from the magnets and there is a complex interaction between them. This results in the the signal only being able to travel to the next port around the circulator. The circulator assembly consisting of the Y junction and the ferrite disks has a distinct resonant frequency - the assembly actually forms a resonator. For obvious reasons the circulator is not operated at this frequency, but either above or below it because the insertion loss, i.e. the attenuation is much lower.

RF Circulator Applications

There are several applications for RF circulators in a variety of RF circuit design applications. Normally they tend to be used at microwave frequencies and as a result they are often referred to as microwave circulators.

Duplexer: One of the most obvious and common applications for an RF circulator is within radar systems or radio communications systems where the transmitter and receiver use a common antenna.

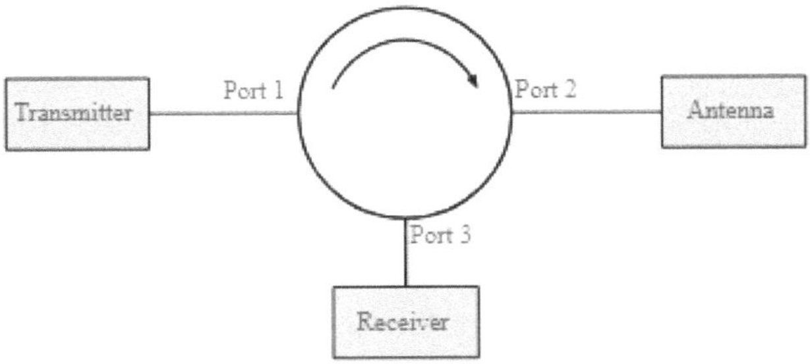

RF circulator used as a duplexer to allow both transmit and receive on a single antenna

RF circulator used as a duplexer

Here the transmitter output is connected for example to Port 1, the antenna to Port 2 and the receiver to Port 3. Accordingly the transmitter power will circulate to the antenna, but not to the receiver, and the antenna received signals will circulate to the receiver. In this way the receiver is isolated from the transmitter but the antenna has power from the transmitter and passes the received signal to the receiver without any mechanical switching.

RF Isolator

The RF circulator can be used as an RF isolator. These are useful for protecting a transmitter output amplifier has to run with a high level of VSWR. As high voltages or current levels can be experienced by the power amplifier if connected directly to the antenna under these circumstances it is possible for the amplifier to be damaged by them. Often transmitters need to operate over a wide bandwidth and under these circumstances, a good impedance match is unlikely to be maintained over the whole bandwidth and damaging levels of VSWR could be seen. To overcome this issue a circulator can be used to protect the PA from the effects of the reflected power.

For this the transmitter is connected to Port 1 and the antenna to Port 2. Port 3 is typically connected to a 50Ω load - this is required because the levels of isolation are dependent upon having a good impedance match presented at the different ports. If a poor match or open circuit is presented, then the isolation performance will be impaired.

RF circulator used as a isolator - note the resistor on Port 3 to provide a matched load for the Port to maintain the level of isolation & dissipate unwanted reflected power.

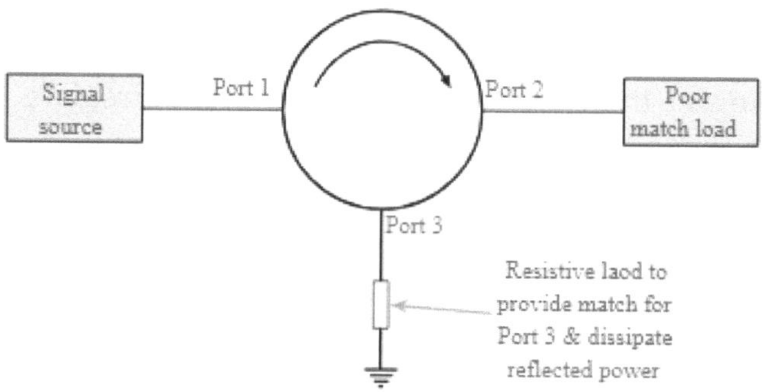

RF Circulator used as An Isolator

The transmitted power is passed from Port 1 to Port 2 and travels to the antenna. Any reflected power will return along the feeder and pass from Port 2 into Port 3 where it can be dissipated in the load. In this way, the RF power amplifier will be able to operate with a high feeder VSWR, but being protected by the circulator acting as an RF isolator.

The load on Port 3 is required to provide a matched impedance: isolators need matched impedances onth e ports to maintain the level of isolation from that port.

A common use of an isolator is shown in Fig. The isolator is connected between a signal generator and some device under test (DUT). If all impedances are matched, the signal passes freely to the DUT. If there is a mismatch at the DUT or if the DUT is disconnected, it creates a high-voltage standing wave ratio (VSWR), causing a large reflected signal. The circulator absorbs this signal, protecting the usually expensive signal generator.

RF Circulator & Isolator Specifications

Understanding the different specifications for an RF circulator enable the correct decisions to be made for a specific RF circuit design or system. Explanations for some of the key specifications appropriate to RF circulators are given below:

Frequency

RF circulators have a limited frequency range over which they operate satisfactorily. The frequency range is limited by the isolation provided between the ports and directions that should not be coupled and the insertion loss that can be achieved.

Typically, RF circulators can operate between frequencies of about 750MHz up to 20 GHz or so, although specialised (and expensive) designs can be made to work on bands between 100 MHz and 100 GHz.

Each RF circulator will be able to operate over a small band. One example on the market now can operate over 3 - 6 GHz, another 7.9 GHz to 8.4 GHz, and so forth. There are naturally many other examples, often designed for specific RF circuit design applications for satellites, radar, etc . .

Impedance

Like all feeders and other electronic components that are included in RF power transfer, a characteristic impedance needs to be specified. The most popular impedance level for RF circulators is 50Ω.

Insertion loss

This is the attenuation measured in dB of the incident signal from one port to the next, i.e. in the forward direction. Typical figures might be in the range from 0.1 to 0.75 dB. The figure will depend upon the actual circulator, its frequency band and other factors.

Generally, the insertion loss of a circulator / isolator loss increases with frequency. Selecting a circulator with the lowest loss will be particularly important where high powers are involved to preserve the transmitted power - at higher powers, there will be more loss in the circulator, and one that can handle the required power is essential as a result.

Isolation

This figure is the attenuation in the reverse flow direction measured n dB. It is typically between about 17 and 35 dB. If greater levels of isolation are needed, then it is possible to place two RF circulators in series.

Power Handling

The power handling specification for RF circulators is particularly important where high power levels are used. As the wiring and ferromagnetic materials will only be able to handle certain power levels it is important these are not exceeded otherwise performance could be sacrificed, or the component could be irreparably damaged. Power levels are typically measured in dBm, dBw or Watts.

Package Type

Circulators are available in a wide variety of packages. Typically they come in manufacturer standard packages with connectors, although other formats including: Drop-In, Surface mount technology, etc are available.

Connectors

For those RF circulators that are connectorised, the type of connector is important. Two of the most popular types of connector are SMA and N-type. SMA is ideal for lower power levels and where space is at a premium, and N-type for higher power levels. It should be remembered that SMA connectors themselves can handle surprisingly high power levels.

VSWR

The VSWR or voltage standing wave ratio indicates how good the match is to the required impedance level. VSWR levels of between 1:1.2 and 1:1.5 are typically achieved.

RF circulators and isolators are widely used in many microwave and other RF circuit design applications. Everything from radio communications equipment to radar, and many other applications. The fact that they transfer power from one port to another specific port only enables them to be used as duplexers allowing a transmitter and receiver to operate at the same time on nearby frequencies, as well as enabling RF power amplifiers to operate into loads that present a poor impedance match.

In view of their specialised nature, RF circulators tend to be bought in as an electronic component, and as a result there are several different electronic component distributors that are able to supply a range of these products.

UNIT - II

SOLID STATE MICROWAVE DEVICES

Introduction- Gunn_Effect Diodes - GaAs Diode - Ridely-Watkins - Hilsun (RWH) Theory - Modes of Operation - Microwave Generation and Amplification - Avalanche transit - Time devices Introduction - Read Diode -IMPATT Diodes - TRAPATT Diodes - BARITT Diodes - Parametric Devices.

2.1 Introduction

Semiconductor devices used for the detection, generation, amplification, and control of electromagnetic radiation with wavelengths from 30 cm to 1 mm (frequencies from 1 to 300 GHz). The number and variety of microwave semiconductor devices, used for wireless and satellite communication and optoelectronics, have increased as new techniques, materials, and concepts have been developed and applied. Passive microwave devices, such as pn and PIN junctions, Schottky barrier diodes, and varactors, are primarily used for detecting, mixing, modulating, or controlling microwave signals. Step-recovery diodes, transistors, tunnel diodes, and transferred electron devices (TEDs) are active microwave devices that generate power or amplify microwave signals.

Typical high-frequency semiconductor materials include silicon (Si), germanium (Ge), and compound semiconductors, such as gallium arsenide (GaAs), indium phosphide (InP), silicon germanium (SiGe), silicon carbide (SiC), and gallium nitride (GaN). In general, the compound semiconductors work best for high-frequency applications due to their higher electron mobilities.

Passive Devices

A PIN (p-type/intrinsic/n-type) diode is a pn diode that has an undoped (intrinsic) region between the p- and n-type regions. The use of an intrinsic region in PIN diodes allows for high-power operation and offers an impedance at microwave frequencies that is controllable by a lower frequency or a direct-current (DC) bias. The PIN diode is one of the most common passive diodes used at microwave frequencies. PIN diodes are used to switch lengths of transmission line, providing digital increments of phase in individual transmission paths, each capable of carrying kilowatts of peak power. PIN diodes come in a variety of packages for microstrip and stripline packages, and are used as microwave switches, modulators, attenuators, limiters, phase shifters, protectors, and other signal control circuit elements.

A Schottky barrier diode (SBD) consists of a rectifying metal-semiconductor barrier typically formed by deposition of a metal layer on a semiconductor. The SBD functions in a similar manner to the antiquated point contact diode and the slower-response pn-junction diode, and is used for signal mixing and detection. The point contact diode consists of a metal whisker in contact with a semiconductor, forming a rectifying junction. The SBD is more rugged and reliable than the point contact diode. The SBD's main advantage over pn diodes is the absence of minority carriers, which limit the response speed in switching applications and the high-frequency performance in mixing and detection applications. SBDs are zero-bias detectors. Frequencies to 40 GHz are available with silicon SBDs, and GaAs SBDs are used for higher-frequency applications.

The variable-reactance (varactor) diode makes use of the change in capacitance of a pn junction or Schottky barrier diode, and is designed to be highly dependent on the applied reverse bias. The capacitance change results from a widening of the depletion layer as the reverse-bias voltage is increased. As variable capacitors, varactor diodes are used in tuned circuits and in voltage-controlled oscillators. For higher-frequency microwave applications, silicon varactors have been replaced with GaAs. Typical applications of varactor diodes are harmonic generation, frequency multiplication, parametric amplification, and electronic tuning. Multipliers are used as local oscillators, low-power transmitters, or transmitter drivers in radar, telemetry, telecommunication, and instrumentation. See Varactor

Active Devices

Transistors are the most widely used active microwave solid-state devices. At very high microwave frequencies, high-frequency effects limit the usefulness of transistors, and two-terminal negative resistance devices, such as transferred-electron devices, avalanche diodes, and tunnel diodes, are sometimes used. Two main categories of transistors are used for microwave applications: bipolar junction transistors (BJTs) and field-effect transistors (FETs). In order to get useful output power at high frequencies, transistors are designed to have a higher periphery-to-area ratio using a simple stripe geometry. The area must be reduced without reducing the periphery, as large area means large interelectrode capacitance. For high-frequency applications the goal is to scale down the size of the device. Narrower widths of the elements within the transistor are the key to superior high-frequency performance. See Transistor

A BJT consists of three doped regions forming two pn junctions. These regions are the emitter, base, and collector in either an npn or pnp arrangement. Silicon npn BJTs have an upper cutoff frequency of about 25 GHz (varies with manufacturing improvements). The cutoff frequency is defined as the frequency at which the current amplification drops to unity as the frequency is raised. The primary limitations to higher frequency are base and emitter resistance, capacitance, and transit time. To operate at microwave frequencies, individual transistor dimensions must be reduced to micrometer or submicrometer size. To maintain current and power capability, various forms of internal paralleling on the chip are used. Three of these geometries are interdigitated fingers that form the emitter and base, the overlaying of emitter and base stripes, and the matrix approach. Silicon BJTs are mainly used in the lower microwave ranges. Their power capability is quite good, but in terms of noise they are inferior to GaAs metal semiconductor field-effect transistors (MESFETs) at frequencies above 1 GHz and are mainly used in power amplifiers and oscillators. They may also be used in small-signal microwave amplifiers when noise performance is not critical.

Heterojunction bipolar transistors (HBTs) have been designed with much higher maximum frequencies than silicon BJTs. HBTs are essentially BJTs that have two or more materials making up the emitter, base, and collector regions (Fig. 1). In HBTs, the major goal is to limit the injection of holes into the emitter by using an emitter material with a larger bandgap than the base. The difference in bandgaps manifests itself as a discontinuity in the conduction band or the valence band, or both. For npn HBTs, a discontinuity in the valence band is required. In general, to make high-quality heterojunctions, the two materials should have matching lattice constants. For very thin layers, lattice matching is not absolutely necessary as the thin layer can be strained to accommodate the crystal lattice of the other material.

Fortunately, the base of a bipolar transistor is designed to be very thin and thus can be made of a strained layer material. Combinations such as AlGaAs/InGaAs and Si/SiGe are possible. See Semiconductor heterostructures

Field-effect transistors (FETs) operate by varying the conductivity of a semiconductor channel through changes in the electric field across the channel. The three basic forms of FETs are the junction FET (JFET), the metal semiconductor FET (MESFET), and the metal oxide semiconductor FET (MOSFET). All FETs have a channel with a source and drain region at each end and a gate located along the channel, which modulates the channel conduction (Fig. 2). Microwave JFETs and MESFETs work by channel depletion. The channel is n-type and the gate is p-type for JFETs and metal for MESFETs. FET structures are well suited for microwave applications because all contacts are on the surface to keep parasitic capacitances small. The cutoff frequency is mainly determined by the transit time of the electrons under the gate; thus short gate lengths (less than 1 μm) are used.

Power devices consist of a number of MESFETs in parallel with air bridges connecting the sources. GaAs MESFET devices are used in low-noise amplifiers (LNAs), Class C amplifiers, oscillators, and monolithic microwave integrated circuits. The performance of a GaAs FET is determined primarily by the gate width and length. The planar structure of a MESFET makes it straightforward to add a second gate which can be used to control the amplification of the transistor. Dual-gate MESFETs can be used as mixers (with conversion gain) and for control purposes. Applications include heterodyne mixers and amplitude modulation of oscillators. See Amplifier, Mixer, Oscillator

The MOSFET has a highly insulating silicon dioxide (SiO2) layer between the semiconductor and the gate; however, silicon MOSFETs are not really considered microwave transistors. Compared with the GaAs MESFET, MOSFETs have lower electron mobility, larger parasitic resistances, and higher noise levels. Also, since the silicon substrate cannot be made semi-insulating, larger parasitic capacitances result. MOSFETs therefore do not perform very well above 1 GHz. Below this frequency, MOSFETs find application mainly as radio-frequency (RF) power amplifiers.

A disadvantage of the MESFET is that the electron mobility is degraded since electrons are scattered by the ionized impurities in the channel. By using a heterojunction consisting of n-type AlGaAs with undopedGaAs, electrons move from the AlGaAs to the GaAs and form a conducting channel at the interface. The electrons are separated from the donors and have the mobility associated with undoped material. A heterojunction transistor made in this fashion has many different names: high electron mobility transistor (HEMT), two-dimensional electron gas FET (TEGFET), modulation-doped FET (MODFET), selectively doped heterojunction transistor (SDHT), and heterojunction FET (HFET). The HEMT has high power gain at frequencies of 100 GHz or higher with low noise levels.

A monolithic microwave integrated circuit (MMIC) can be made using silicon or GaAs technology with either BJTs or FETs. For high-frequency applications, GaAs FETs are the best choice. A MMIC has both the active and passive devices fabricated directly on the substrate. MMICs are typically used as low-noise amplifiers, as mixers, as modulators, in frequency conversion, in phase detection, and as gain block amplifiers. Silicon MMIC devices operate in the 100-MHz to 3-GHz frequency range. GaAs FET MMICs are typically used in applications above 1 GHz.

Active Microwave Diodes

Active microwave diodes differ from passive diodes in that they are used as signal sources to generate or amplify microwave frequencies. These include step-recovery, tunnel, Gunn, avalanche, and transit time diodes, such as impact avalanche and transit-time (IMPATT), trapped plasma avalanche triggered transit-time (TRAPATT), barrier injection transit-time (BARITT), and quantum well injection transit time (QWITT) diodes.

A step recovery diode is a special PIN type in which charge storage is used to produce oscillations. When a diode is switched from forward to reverse bias, it remains conducting until the stored charge has been removed by recombination or by the electric field. A step recovery diode is designed to sweep out the carriers by an electric field before any appreciable recombination has taken place. Thus, the transition from the conducting to the nonconducting state is very fast, on the order of picoseconds. Because of the abrupt step, this current is rich in harmonics, so these diodes can be used in frequency multipliers.

For microwave power generation or amplification, a negative differential resistance (NDR) characteristic at microwave frequencies is necessary. NDR is a phenomenon that occurs when the voltage (V) and current (I) are 180° out of phase. NDR is a dynamic property occurring only under actual circuit conditions; it is not static and cannot be measured with an ohmmeter. Transferred electron devices (TEDs), such as Gunn diodes, and avalanche transit-time devices use NDR for microwave oscillation and amplification. TEDs and avalanche transit-time devices today are among the most important classes of microwave solid-state devices. See Negative-resistance circuits

The tunnel diode uses a heavily doped abrupt pn junction resulting in an extremely narrow junction that allows electrons to tunnel through the potential barrier at near-zero applied voltage. This results in a dip in the current-voltage (I-V) characteristic, which produces NDR. Because this is a majority-carrier effect, the tunnel diode is very fast, permitting response in the millimeter-wave region. Tunnel diodes produce relatively low power. The tunnel diode was the first semiconductor device type found to have NDR. See Tunnel diode

Avalanche diodes are junction devices that produce a negative resistance by appropriately combining impact avalanche breakdown and charge-carrier transit time effects. Avalanche breakdown in semiconductors occurs if the electric field is high enough for the charge carriers to acquire sufficient energy from the field to create electron-hole pairs by impact ionization. The avalanche diode is a pn-junction diode reverse-biased into the avalanche region. By setting the DC bias near the avalanche threshold, and superimposing on this an alternating voltage, the diode will swing into avalanche conditions during alternate half-cycles. The hole-electron pairs generated as a result of avalanche action make up the current, with the holes moving into the p region, and the electrons into the n region. The carriers have a relatively large distance to travel through the depletion region. At high frequencies, where the total time lag for the current is comparable with the period of the voltage, the current pulse will lag the voltage. By making the drift time of the electrons in the depletion region equal to one-half the period of the voltage, the current will be 180° out of phase. This shift in phase of the current with respect to the voltage produces NDR, so that the diode will undergo oscillations when placed in a resonant circuit.

A Gunn diode is typically an n-type compound semiconductor, such as GaAs or InP, which has a conduction band structure that supports negative differential mobility. Although this device is referred to as a Gunn diode, after its inventor, the device does not contain a pn junction and can be viewed as a resistor below the threshold electric field (Ethres). For applied voltages that produce electric fields below Ethres, the electron velocity increases as the electric field increases according to Ohm's law. For applied voltages that produce electric fields above Ethres, conduction band electrons transfer from a region of high mobility to low mobility, hence the general name –transferred electron device.'' Beyond Ethres, the velocity suddenly slows down due to the significant electron transfer to a lower mobility band producing NDR. For GaAs, Ethres is about 3 kV/cm. The Gunn effect can be used up to about 80 GHz for GaAs and 160 GHz for InP. Two modes of operation are common: nonresonant bulk (transit-time) and resonant limited space-charge accumulation (LSA).

Impact avalanche and transit-time diodes (IMPATTs) are NDR devices that operate by a combination of carrier injection and transit time effects. There are several versions of IMPATT diodes, including simple reverse-biased pn diodes, complicated reverse-biased multidopedpn layered diodes, and reverse-biased PIN diodes. The IMPATT must be connected to a resonant circuit. At bias turn-on, noise excites the tuned circuit into a natural oscillation frequency. This voltage adds algebraically across the diode's reverse-bias voltage. Near the peak positive half-cycle, the diode experiences impact avalanche breakdown. When the voltage falls below this peak value, avalanche breakdown ceases. A 90° shift occurs between the current pulse and the applied voltage in the avalanche process. A further 90° shift occurs during the transit time, for a total 180° shift which produces

NDR. An IMPATT oscillator has higher output power than a Gunn equivalent. However, the Gunn oscillator is relatively noise-free, while the IMPATT is noisy due to avalanche breakdown.

A trapped plasma avalanche triggered transit-time (TRAPATT) diode is basically a modified IMPATT diode in which the holes and electrons created by impact avalanche ionization multiplication do not completely exit from the transit domain of the diode during the negative half-cycle of the microwave signal. These holes and electrons form a plasma which is trapped in the diode and participates in producing a large microwave current during the positive half-cycle.

A barrier injection transit-time diode (BARRITT) is basically an IMPATT structure that employs a Schottky barrier formed by a metal semiconductor contact instead of a pn junction to create similar avalanche electron injection.

A variety of approaches have been investigated to find alternative methods for injecting carriers into the drift region without relying on the avalanche mechanism, which is inherently noisy. Quantum well injection transit-time diodes (QWITT) employ resonant tunneling through a quantum well to inject electrons into the drift region. The device structure consists of a single GaAs quantum well located between two AlGaAs barriers in series with a drift region of made of undopedGaAs. This structure is then placed between two n+-GaAs regions to form contacts.

2.2 Gunn - Effect Diodes

What is Gunn Diode: Construction & Its Working

In GaAs semiconductor materials, the electrons are present in two states like high mass low velocity & low mass high velocity. By the demand of an adequate electric field, the electrons are forced to move from a low mass state to a high mass state. At this specific state, electrons can form a group & moves at a consistent rate that can cause to flow current in a series of pulses. So this is known as Gunn Effect which is used by Gunn diodes. These diodes are the best and most frequently available devices from the TED family (transferred electron devices). These types of diodes are used like DC to microwave converters with the negative resistance features of bulk GaAs (Gallium Arsenide) and they need a typical, stable voltage power supply, less impedance so that complex circuitry can be eliminated. This article discusses an overview of a Gunn diode.

What is a Gunn Diode?

Gunn diode is made with N-type semiconductor because it comprises majority charge carriers like electrons. This diode uses the negative resistance property to produce current at high frequencies. This diode is mainly used to produce microwave signals around 1 GHz & RF frequencies around 100 GHz.

Gunn diodes are also known as TED (transferred electron devices). Even though it is a diode, the devices do not have a PN-junction but include an effect called the Gunn Effect.

Gunn Diode

This effect was named based on the inventor namely J B Gunn. These diodes are very simple to use, they form a low-cost technique to generate microwave RF signals, frequently being placed in a waveguide to make an easy resonant cavity. The Gunn diode symbol is shown below.

Symbol

Gunn Diode Construction

The fabrication of the Gunn diode can be done with an N-type semiconductor. The materials which are used most frequently are GaAs (gallium Arsenide) &InP (Indium Phosphide) and other materials have been utilized like Ge, ZnSe, InAs, CdTe, InSb.

It is essential to utilize n-type material because the effect of the transferred electron is simply appropriate to electrons & not holes found in a p-type material. In this device, there are 3 main regions which are called top, bottom & middle areas.

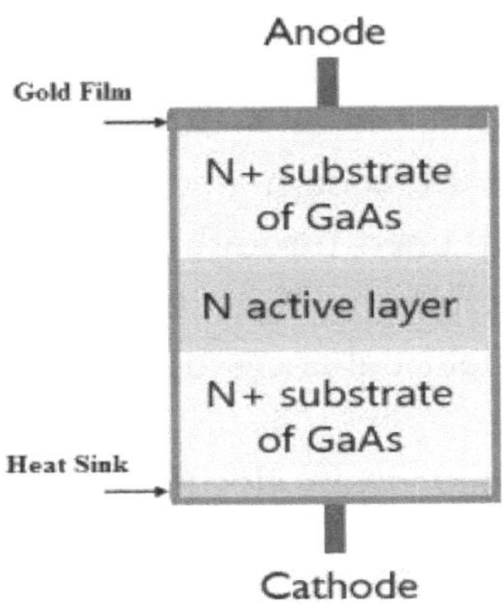

Construction

The general method to manufacture this diode is to grow & epitaxial layer on a degenerate n+ substrate. The thickness of the active layer ranges from few microns to 100 microns and the doping level of this layer ranges from 1014cm-3 to 1016cm-3. But this doping level is significantly low which is used for the top & bottom regions of the device. Based on the required frequency, the thickness will change.

The deposition of the n+ layer can be done epitaxially otherwise doped through ion implantation. Both the areas of this device like top & bottom are doped deeply to provide n+ material. This gives the necessary high conductivity regions that are required for the connections toward the device.

Generally, these devices are placed on conducting support to which the connection of a wire is made. This support can also work like a heat sink which is dangerous to remove the heat. The diode's other terminal connection can be made through a gold connection that is deposited onto the pinnacle surface. Here the gold connection is necessary due to its high conductivity & relative stability. While manufacturing, the material device should be defect-free & also include an extremely consistent range of doping.

Gunn Diode Working

The working principle of a Gunn diode mainly depends on the Gunn Effect. In some materials like InP&GaAs, once a threshold level is attained through an electric field within the material, then electrons mobility will decrease concurrently. When the electric field enhances then negative resistance will be generated.

Once the intensity of an electric field for GaAs material reaches its significant value on the negative electrode, then low electron mobility region can be formed. This region moves through the average electrons speed to the +Ve electrode.

Gunn diode includes a negative resistance region on its CV characteristics. Once the significant value is attained through the negative GaAs electrode, then there will be a region through the mobility of low electrons. After that, it will shift to the positive electrode. Once it meets a strong electric field domain through the positive electrode on the negative electrode, then a cyclic type of the region for less electron mobility as well as the high electric field will begin to re-create.

The cyclical nature of this incident produces oscillations with 100 GHz frequencies. Once this value exceeds, then oscillations will begin to disappear quickly.

Characteristics

The Gunn diode characteristics show a negative resistance area on its VI characteristic curve shown below. So this region allows the diode to amplify signals, so it can be used in oscillators & amplifiers. But, the Gunn diode oscillators are used most frequently.

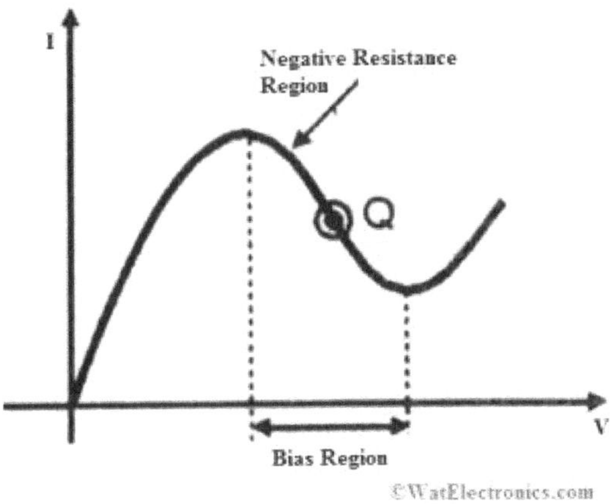

Characteristics of Gunn Diode

Here, the negative resistance area in the Gunn diode is nothing but once the flow of current increases then the voltage drops. This phase reverse allows the diode to work like an oscillator & an amplifier.

The flow of current in this diode increases through the DC voltage. At a specific end, the flow of current will start decreasing, so this is called a peak point or threshold point. Once the threshold point is crossed then the flow of current will start reducing to create a negative resistance region within the diode.

Advantages

The advantages of the Gunn diode include the following.

- This diode is available in small size & portable
- Making the cost of this diode is less
- At high frequencies, this diode is stable & reliable
- It possesses an improved noise-to-signal ratio (NSR) because it is protected from noise annoyance.
- It includes high bandwidth

Disadvantages

The disadvantages of the Gunn diode include the following.

- The temperature stability of this diode is poor
- The operating current of this device, thus power dissipation is high.
- The Gunn diode efficiency is low under10GHz.
- Turn on the voltage of this device is high
- FM noise is high for specific applications
- The range of tuning is high

Applications

The applications of Gunn diode include the following.

- These diodes are used as oscillators & Amplifiers.
- It is used in microelectronics like control equipment.
- These are used in military, commercial radar sources & radio communication.
- This diode is used in pulsed Gunn diode generators.
- In microelectronics, these diodes are utilized as quick controlling devices for laser beam modulation.
- Used in police radars.
- These diodes are applicable in tachometers
- It is used as pump sources within parametric amplifiers
- Used in sensors to detect different systems like a door opening, trespass detecting & pedestrian safety, etc.
- It is used in nonstop wave doppler radars.
- It is used widely in transmitters of microwave relay data link
- It is used in electronics oscillators for generating microwave frequencies

2.3 GaAs Diode

Gallium arsenide is the orginal microwave semiconductor that helped spawn the revolution in personal communications that we all take for granted. But now GaAs is an older gentleman, soon to be content spending days on the park bench of microwave power amplifiers, watching gallium nitride take over this end of the business...

- GaAs MESFET
- GaAs PHEMT
- GaAs MHEMT
- GaAs HBT
- GaAs VPIN diode

GaAs MESFET

Gallium arsenide MESFET was the original answer to "how can we make amplifiers at microwave frequencies?" The first GaAs MMICs demonstrated in the 1970s. Including HEMT and HBT technologies, literally billions of dollars have been spent extending fmax of GaAs products up into 100s of GHz.

The semi-insulating properties of GaAs substrates and the 12.9 dielectric constant make it an EXCELLENT media for microstrip or CPW design. GaAs devices operate reliably up to 150C channel temperature, although some suppliers will tell you it is good for 175C or even higher. It is "radiation hard" for space applications. GaAs substrates are available up to six inches (150 mm) in diameter, which has been a long development since the first 2-inch wafers were available in the late 1970s. Sadly, GaAs MESFET MMICs will NEVER be cheaper than silicon, due to the starting material cost ($100s of dollars). GaAs parts are more fragile than silicon, and the thermal dissipation factor is not that good. GaAs MESFETs may be extinct in five years, because it doesn't cost much more to fabricate PHEMT or MHEMT on GaAs, and these technologies offer higher performance.

Advantages

- Mature technology
- Optical gates (usually) means low cost
- Great microwave substrate (12.9 Er, low loss tangent, high bulk resistivity)
- Six inch wafers available
- Photonic properties
- 16-20 volt breakdown possible
- Relatively cheap to produce (but always more than silicon)
- Channel temperatures up to 150C possible

Disadvantages

- Limited to Ku-band or lower
- Noise figure and power performance not as good as GaAs PHEMT
- Positive and negative voltage typically needed (VGS and VDS)

Examples

M/A-COM Roanoke foundry (to be moved in 2010), TriQuint Oregon

GaAs PHEMT

GaAs PHEMT was the second MMIC technology to be perfected, in the 1990s. Breakdown voltages of PHEMT up to 16 volts make high-power/high efficiency amps possible, and noise figure of tenths of a dB at X-band means great LNAs, and made the DISH network possible, you lucky dogs!
PHEMT stands for pseudomorphic high electron mobility transistor. "Pseudomorphic" implies that the semiconductor is not just GaAs, perhaps AlGaAs/InGaAs/GaAs or some other secret recipe of 11 herbs and spices. Here's some further info on the the use of pseudomorphic in this context (sent in by some M101 fans!)

Actually, "pseudomorphic" means that the hetero layers are thin enough not to keep their own crystal lattice structure, but assume the structure (lattice constants especially) of surrounding material (lots of stress is involved),

If you look at a two dimensional cross section of the layer, you'll see that while it assumes the lattice constant of the bulk structure in the X direction, it tries to keep its original lattice constant in the vertical direction. This layer is indeed strained. For a GaAspHEMT, indium is added to improve mobility and form a quantum well. Indium wants to growth the lattice and the typical range for useful thicknesses would be 10-25% on GaAs. You can also do strain compensation with the Schottky or cap layer.

The purist nerds of semiconductors often capitalize "PHEMT" as pHEMT. To them we offer this advice: get over it, or we will beat you up like we used to do on the playground, remember?

Advantages

- Useful through Q-band, especially if thinned to 2 mils and individual source vias are used
- Excellent power and efficiency (greater than 60% PAE)
- Breakdown 12 volts at best, typical operate at 5-6 volts
- Channel temperatures up to 150C possible

Disadvantages

- E-beam gates (increases cost)
- Positive and negative voltage typically needed (VGS and VDS)

Examples:

TriQuint Texas, Northrop, Raytheon (in general all foundries catering to military products)

GaAs MHEMT

Recent work on metamorphic MHEMT has made premium InP HEMT performance possible (amps up at 100 GHz) at the same price as "regular" GaAs PHEMT. You can get noise figure and fmax equal to indium phosphide by using MHEMT, if you use a reputable foundry and indium content is high. You can actually exceed InP RF performance with indium content greater than 55%! The down side to all that indium is reduced operating voltage.

MHEMT stands for metamorphic high-electron mobility transistor. The channel material is InGaAs. "Metamorphic" implies that the lattice structure of GaAs is buffered using epitaxial layers to gradually transform the lattice constant so it lines up with InGaAs. InGaAs is normally grown on InP, which is expensive and fragile compared to GaAs. "Metamorphic" is changing the lattice constant by bond breaking as opposed to "pseudomorphic" which means just straining the heck out of it!

Advantages

- Extremely low noise figure
- Incredibly high fmax (more than 100 GHz)
- Extremely low on-resistance, makes great switches, but not as good as PIN diodes
- Channel temperatures up to 150C possible

Disadvantages

- Breakdown voltage much lower than PHEMT
- Low operating voltage (1 to 2 volts)
- Positive and negative voltage typically needed (VGS and VDS)

Examples

BAE, Win Semi

GaAs HBT

The heterojunction bipolar transistor (HBT) is a new development, and can decrease the cost of GaAs amplifier products because the emitters are formed optically. GaAs HBT devices operate vertically, compared to the horizontal operation of FETs. However, for very high frequency, the emitter size must be made quite small, and the InGaAs layer is thick and is a thermal insulator, so these devices tend to run HOT. Typical HBT amps are "gain blocks", used in the UHF to C-band frequency ranges.

Typical supporters of HBTs will tell you that wafer yield up to 99% is possible.

Advantages

- Single power supply polarity
- All-optical process

Disadvantages

- Heat dissipation can be problem at small emitter size
- Typically, reverse isolation is not as high as with PHEMT amplifiers, leading to poor amplifier directivity
- Collector resistors are required to stabilize amplifiers. These cut into your power efficiency

Examples

RFMD

GaAs VPIN Diode

PIN diodes make great switching elements. Vertical PINs (VPINs) are offered on some MMICs, but this is truly a niche market. As far as we know, nobody offers VPIN diodes and amplifier devices such as FETs on the same wafer.

Advantages

- The lowest on-resistance for the least amount of off-capacitance.
- Huge power handling.

Disadvantages

- Two terminal device means you must create bias tees to bring in DC control signals.
- Expect DC current up to 20 mA to create a good RF short circuit.

Examples

M/A-COM (Cobham), TriQuint Texas

2.4 Ridely - Watkins - Hilsun (RWH) Theory

RWH Theory or Two-valley Theory of Gunn Diode

The Gunn diode, which is made of n-doped semiconductor material (e.g. GaAs or InP), is characterized by having two valleys in their conduction bands with different mobility. The **two-valley model** is also called the **Ridley–Watkins–Hilsum (RWH) theory**. There are two regions in the conduction band of the N-type GaAs. These conduction band regions are known as an upper valley and lower valley. The energy band and transfer of electrons between the two valleys are shown in Figures 9.4 and 9.5, respectively.

Key points of GaAs

- In N-type GaAs, the valence band is filled with electrons, and the conduction band is partly filled.
- The forbidden energy gap between the valence band and the conduction band is about 1.43 eV.
- Electrons in the lower valley (n_1) exhibit a small effective mass (m_1) and very high mobility, μ_1.
- Electrons in the upper valley (n_2) exhibit a large effective mass (m_2) and very low mobility, μ_2.
- The two valleys are separated by a small energy gap, ΔE, of approximately 0.36 eV. At a low electric field, electrons remain in the lower valley and material

behaves ohmically. When the electric field reaches a certain threshold value E_{th}, the electrons will be swept from the lower valley to the upper valley. If the rate of transfer of electrons from the lower to the upper valley is very high, the current will decrease with an increase in voltage.

- The average electron mobility, μ is given by
- $\mu=(n1\mu1+n2\mu2)/(n1+n2)$
- Where,
- n_1 = electron density in lower valley
- n_2 = electron density in upper valley
- μ_1 = mobility of electron in lower valley
- μ_2 = mobility of electron in upper valley
- Total carrier concentration = $n_1 + n_2$

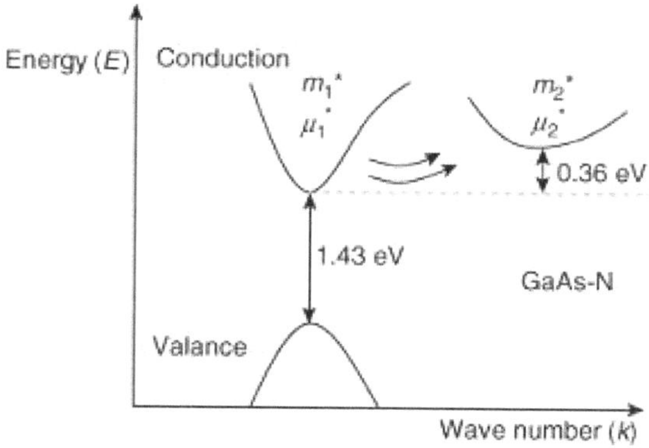

Energy band diagram for GaAs

Where,

m_1^* = effective mass of electron in lower valley
m_2^* = effective mass of electron in upper valley

Typical Characteristics

The voltage of operation depends on the active layer width. As the device has to operate between Eth (3.3 kV/cm) and Emin (11 kV/cm) say at E0 = 6 kV/cm. The current density at 6 kV/cm is 400 A/cm^2.

Therefore requirement of bias depends on active layer width, e.g. for L = 5 μm, V = 3 V and for L = 50 μm, V = 30 V. Therefore for 3×10^{-8} m^2 area, I = 120 mA. With the above, the characteristic of a typical diode may be as:

- CW power = 25 mW to 250 mW X band (5 – 15 GHz)
- 100 mW at 18-26.5 GHz
- 40 mW at 26.5-40 GHz
- Pulsed Power = 5W (5-12 GHz)
- Efficiency = 2–12%
- Frequency Tuning: The frequency of Gunn oscillator can be changed by mechanical tuning and electronic tuning.

Applications of Gunn Diode

1. It is used in Radar transmitters as a low power oscillator (e.g. CW Doppler Radar, Police Radar).
2. Gunn diode oscillators with pulse signals are used in transponders in industry telemetry systems and for air traffic control.
3. It is used in a broadband linear amplifier.
4. Gunn diode is used in sequential logic circuits and fast combinational circuits.
5. It is used in microwave receivers as low and medium power oscillator.
6. It is used in a parametric amplifier as pump sources.

2.5 Modes of Operation

The operation of a Gunn diode can be done in four modes which include the following.

- Gunn Oscillation Mode
- Stable Amplification Mode

- LSA Oscillation Mode
- Bias Circuit Oscillation Mode
- Gunn Oscillation Mode

Gunn oscillation mode can be defined in the area wherever the sum of frequency can be multiplied by 107 cm/s lengths. The sum of doping can be multiplied through the length is higher than 1012/cm2. In this region, the diode is not stable due to the formation of cyclic either the high field domain & the accumulation layer.

Stable Amplification Mode

This kind of mode can be defined in the area wherever the sum of frequency times length is 107cm/sec & the doping product length for time ranges from 1011 & 1012/cm2.

LSA Oscillation Mode

This kind of mode can be defined in the area where the sum of times length of frequency is 107 cm/s & the doping quotient can be divided through the frequency is ranges from 2×104 & 2×105.

Bias Circuit Oscillation Mode

This kind of mode happens simply once there is either LSA or Gunn oscillation takes place. Generally, it is the area wherever the time's length product of frequency is very small to appear within the figure. Once the biasing of a bulk diode is done to the threshold then the average current drops suddenly when the oscillation of Gunn starts.

Gunn Diode Oscillator Circuit

The circuit diagram of the Gunn diode oscillator circuit is shown below. The application of the Gunn diode diagram shows a negative resistance region. The negative resistance through stray capacitance and lead inductance can result in oscillations.

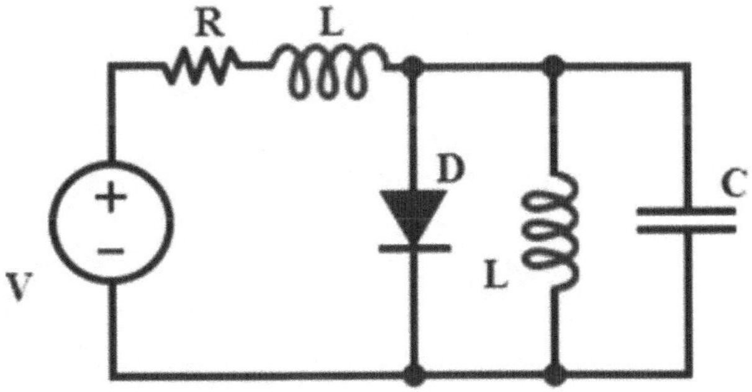

Gunn Diode Oscillator Circuit

In most cases, the relaxation kind of oscillations will include huge amplitude which will damage the diode. So a large capacitor is used across the diode to avoid this failure. This characteristic is mainly used to design oscillators at upper frequencies that range from GHz to THz bands. Here, the frequency can be controlled by adding a resonator. In the above circuit, the lumped circuit equivalent is a waveguide or coaxial transmission line.

Here, GaAs Gunn diodes are accessible for operation which ranges from 10 GHz – 200 GHz at 5 MW – 65 MW power. These diodes can also be used as amplifiers.

2.6 Microwave Generation and Amplification

This is a (virtually) math-free introduction to microwave amplifiers from an applications standpoint. There are many references available for the aspiring amplifier designer; this series of posts will attempt to quickly elucidate the relevant factors for the RF system design engineer working to evaluate the appropriate amplifier for her system design.

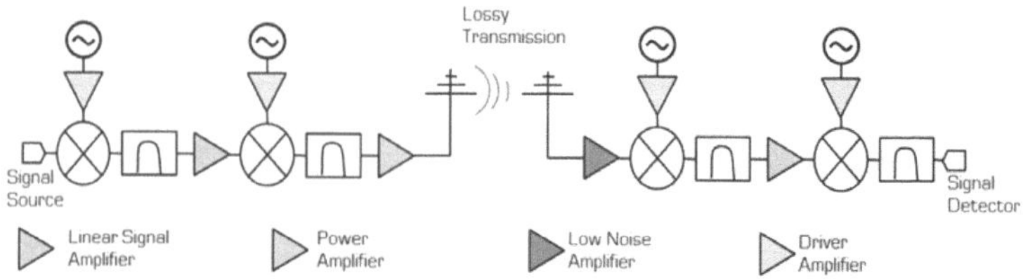

Figure: Types of Microwave Amplifiers

There are many ways to classify microwave amplifiers, but we will group them into four categories based on what role they would play in a generic superheterodyne receiver (shown above). The above system could represent a cellular, backhaul, satellite, or other radio communications link; it could also represent a radar or other imaging system. The transmitter alone could represent a jammer or exciter, and the receiver alone could represent an electronic warfare scanner or a test instrument. The requirements that drive the amplifier selection will be the same for most applications.

We will discuss the different types of amplifiers and what they are trying to accomplish, starting with the

Low Noise Amplifier (LNA)

- Critical Figures of Merit:
- Gain

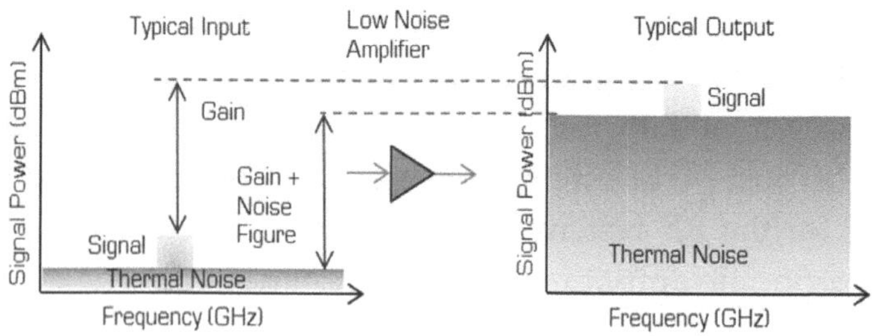

Figure: Noise Figure/Noise Factor/Noise Temperature

The low noise amplifier is the most straightforward microwave amplifier to understand due the relative insignificance of nonlinear effects in the amplifier due to the low input power. It is designed to take a very low level signal such as you would see at the end of a lossy transmission medium and amplify it with minimal additive noise. It is critical to note that the noise is amplified with the same gain as the signal, in addition to the additive noise. This means that a low noise amplifier or any other amplifier cannot increase the signal to noise ratio (SNR), it can only raise the power level of both the signal and the noise. If the gain is narrowband, the amplifier may filter some out of band noise or signals, but it cannot improve the signal to noise ratio due to the in-band noise.

If the SNR is not improved, then why bother amplifying? The reason is that successive stages in the receiver have loss, and that loss is also equally applied to the signal and noise. A loss will not significantly degrade the SNR of the signal if the noise level is already well above the noise floor. Therefore, the LNA noise figure typically dominates the noise figure for the entire receiver chain. Finally, the signal detector at the end of the signal chain (typically an analog to digital converter (ADC) in modern systems) has a detectable voltage range, and the receiver designer will use the amplifier stages throughout the receiver chain to match the received signal power to the detectable voltage range of the ADC.

Low noise amplifiers are typically narrowband, or at least band limited, devices. This is because, as we will see in a future blog post about amplifier circuits, tradeoffs must be made to reduce the noise figure of the amplifier. Typically this means creating a narrowband matching network that optimizes the noise figure of the transistor in the amplifier. Modern LNAs can achieve extremely high gains with low noise figures, as much as 40 dB of gain with less than 1 dB of noise figure at low frequencies.

Power Amplifier (PA)
Critical Figures of Merit

- Output Power (Psat and P1dB)
- Efficiency
- Nonlinearity (IP2, IP3, Spectral Regrowth)

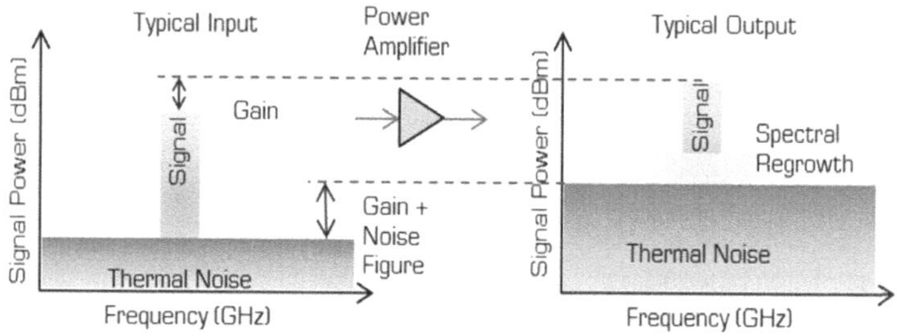

Figure: Power Amplifier Operation

The power amplifier is the opposite of the low noise amplifier. A power amplifier takes a signal that is already at a relatively high level and boosts it for transmission over a lossy medium such as through the air. While the gain of a PA is typically low (less than 10 dB), the power is significant in absolute terms (i.e. an input of 500 milliwatts could be boosted to several watts). The saturated output power (Psat) of the PA largely determines the physical range in terms of achievable transmission/detection distance of the wireless system.

Saturated output power is obviously the most important parameter of a power amplifier, and efficiency is very closely related. Efficiency determines battery life for portable applications, transmission power for airborne/satellite applications, and heat dissipation requirements for all applications. Most of the power in many transceivers is used in the final power amplifier.

Since both the signal and the noise output from the PA will be heavily attenuated by the transmission medium, the noise figure is not critical. Older frequency/phase only transmission formats like frequency modulation (FM), frequency shift keying (FSK), binary phase shift keying (BPSK) and others used signals with constant amplitudes, meaning that power amplifier nonlinearity was not a factor in signal quality. This was necessary since the tube amplifiers of the time had strong nonlinearities. Modern communication formats such as Quadrature Amplitude Modulaton (QAM), Orthogonal Frequency Division Multiplexing (OFDM), Code Division Multiple Access (CDMA), and most others use both amplitude and phase modulation and therefore require linear amplification. Many applications also require simultaneous amplification of multiple unrelated channels.

This nonlinear behavior of the amplifier is described in many ways, and ultimately manifests as both degraded signal quality and legally restricted signal leakage into adjacent channels, as described by the Adjacent Channel Power Ratio (ACPR). These effects are often exacerbated by demands that the amplifier simultaneously amplify signals with large, unpredictable peaks in power. This signal characteristic is described by the signal peak to average power ratio (PAPR), and modern signals have much higher PAPR than their predecessors.

By constantly demanding more information be transmitted in the same limited bandwidth, the cellular/wireless revolution has required linear amplification of high PAPR signals without leakage into adjacent channels and without draining the battery or creating too much heat. The billions of dollars that chased this demand led to an explosion of research and innovation in power amplifier circuits that we will briefly review in a later post.

Linear Signal Amplifier

Critical Figures of Merit:

- Nonlinearity (IP2, IP3, Spectral Regrowth)
- Noise Figure
- Usability

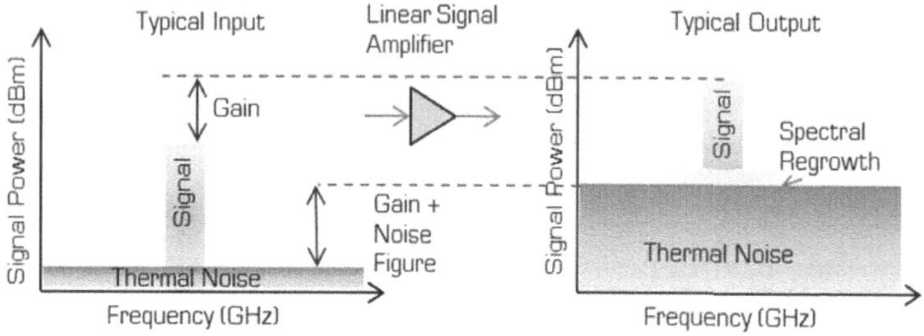

Figure: Linear Amplifier Operation

Linear signal amplifiers are generic amplifiers, sometimes called ‗gain blocks', that provide signal gain within a system. Since they are not at the input or the output of the system, they are generally not the determining factor in the dynamic range of the system. Power levels are higher than into the LNA, but lower than the PA, and as such the choice of amplifier is a balance of linearity, noise figure, and what we will call ‗convenience factors' such as cost, size, and power consumption. While the requirements placed on them are not extreme, they must not degrade the signal in any significant way. These amplifiers are never the star of the system, but the team still loses if they don't do their job.

Driver Amplifier

Critical Figures of Merit:

- Output Power (Psat)
- Phase Noise
- Efficiency
- Harmonic Generation

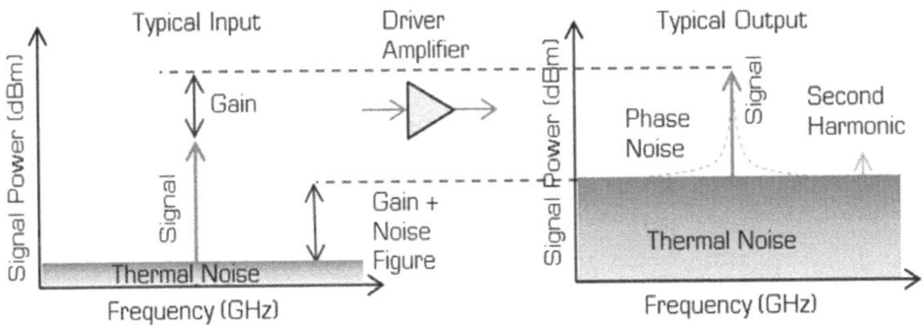

Figure: Driver Amplifier Operation

The driver amplifier is the underserved application space that Marki specializes in. These amplifiers are specifically used for single frequency (continuous wave or CW) operation as in synthesizers or as amplifiers for the local oscillator (LO) driving a mixer. In consumer applications the role of the driver amplifier is typically integrated into the transceiver chip as a CMOS circuit. In high end electronic warfare, instrumentation, and test and measurement applications, however, these amplifiers are discrete to provide maximum system performance.

Since these amplifiers are specifically used for amplifying a CW signal, the nonlinearity of the amplifier in terms of multitone intermodulation is not relevant; multitone inputs are assumed not to be present. Instead the important manifestation of nonlinearity is harmonic generation. In some applications harmonics are critical to reduce or filter before using the amplifier to drive a mixer because they can degrade the spurious performance of the mixer. In other circumstances (as with the T3), the harmonics actually improve the spurious performance. This will be covered in depth in a future post. In general a driver amplifier with less than an octave of bandwidth can be run near Psat if the harmonics are filtered since all distortion products will be at 2f or higher frequencies.

As with multi-tone intermodulation, the noise figure of a driver amplifier is not critical as long as the noise of the signal source is not too high and the noise rejection of the subsequent circuitry (such as the LO-IF or LO-RF isolation of a mixer) is high enough that it eliminates the amplitude noise. In contrast to the amplitude noise, any phase noise added to the signal (or amplitude noise to phase noise conversion that occurs) in the amplifier will directly degrade the signal integrity of either radar signals or phase controlled communications signals. This can be a serious problem for modern dense constellation QAM signals in particular. Similar to harmonic content, the phase noise of commercial amplifiers is almost never specified.

Since these amplifiers are required to generate LO signals that are always at least 6-10 dB higher power than the signal itself, they are required to be higher power amplifiers than the linear signal amplifiers in the system. Similarly, since there are several of these amplifiers in a super-heterodyne system, especially a multi-channel superheterodyne system, they can be the largest consumer of power in a receiver. Therefore efficiency is very important for driver amplifiers.

2.7 Avalanche Transit Time Devices Introduction

The process of having a delay between voltage and current, in avalanche together with transit time, through the material is said to be Negative resistance. The devices that helps to make a diode exhibit this property are called as Avalanche transit time devices.

The examples of the devices that come under this category are IMPATT, TRAPATT and BARITT diodes. Let us take a look at each of them, in detail.

2.8 IMPATT Diode

This is a high-power semiconductor diode, used in high frequency microwave applications. The full form IMPATT is IMPact ionization Avalanche Transit Time diode.

A voltage gradient when applied to the IMPATT diode, results in a high current. A normal diode will eventually breakdown by this. However, IMPATT diode is developed to withstand all this. A high potential gradient is applied to back bias the diode and hence minority carriers flow across the junction.

Application of a RF AC voltage if superimposed on a high DC voltage, the increased velocity of holes and electrons results in additional holes and electrons by thrashing them out of the crystal structure by Impact ionization. If the original DC field applied was at the threshold of developing this situation, then it leads to the avalanche current multiplication and this process continues. This can be understood by the following figure.

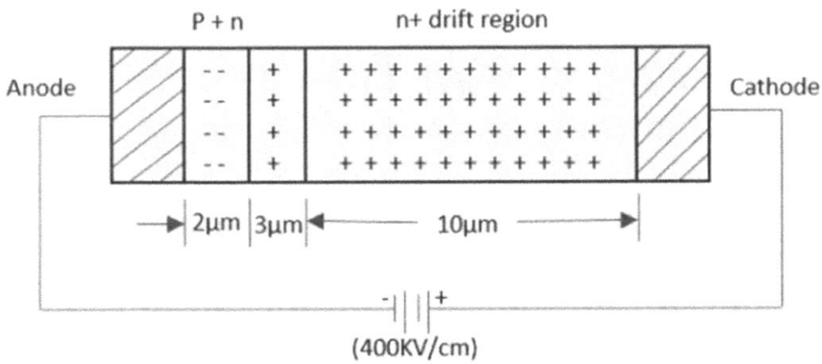

IMPATT diode schematic

Due to this effect, the current pulse takes a phase shift of 90°. However, instead of being there, it moves towards cathode due to the reverse bias applied. The time taken for the pulse to reach cathode depends upon the thickness of n+ layer, which is adjusted to make it 90° phase shift.

Now, a dynamic RF negative resistance is proved to exist. Hence, IMPATT diode acts both as an oscillator and an amplifier.

The following figure shows the constructional details of an IMPATT diode.

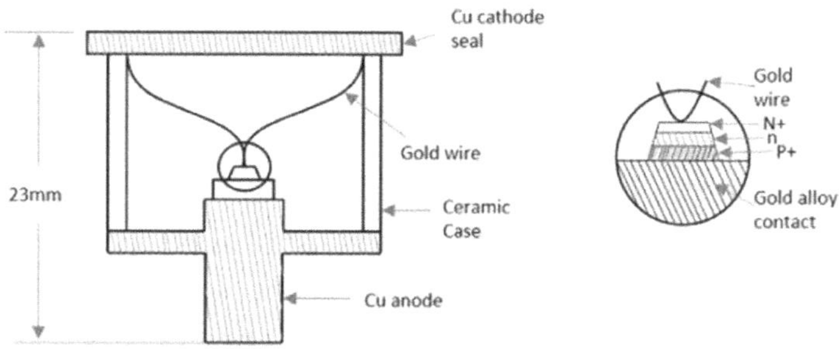

Constructional details of IMPATT

The efficiency of IMPATT diode is represented as

$$\eta = \left[\frac{P_{ac}}{P_{dc}}\right] = \frac{V_a}{V_d}\left[\frac{I_a}{I_d}\right]$$

Where,

- P_{ac} = ACpower
- P_{dc} = DCpower
- V_a&I_a= AC voltage & current
- V_d&I_d = DC voltage & current

Disadvantages

Following are the disadvantages of IMPATT diode.

- It is noisy as avalanche is a noisy process
- Tuning range is not as good as in Gunn diodes

Applications

Following are the applications of IMPATT diode.

- Microwave oscillator
- Microwave generators
- Modulated output oscillator
- Receiver local oscillator
- Negative resistance amplifications
- Intrusion alarm networks highQIMPATT
- Police radar highQIMPATT
- Low power microwave transmitter highQIMPATT
- FM telecom transmitter lowQIMPATT
- CW Doppler radar transmitter lowQIMPATT

2.9 Trapatt Diode

The full form of TRAPATT diode is TRApped Plasma Avalanche Triggered Transit diode. A microwave generator which operates between hundreds of MHz to GHz. These are high peak power diodes usually n+- p-p+ or p+-n-n+ structures with n-type depletion region, width varying from 2.5 to 1.25 Âμm. The following figure depicts this.

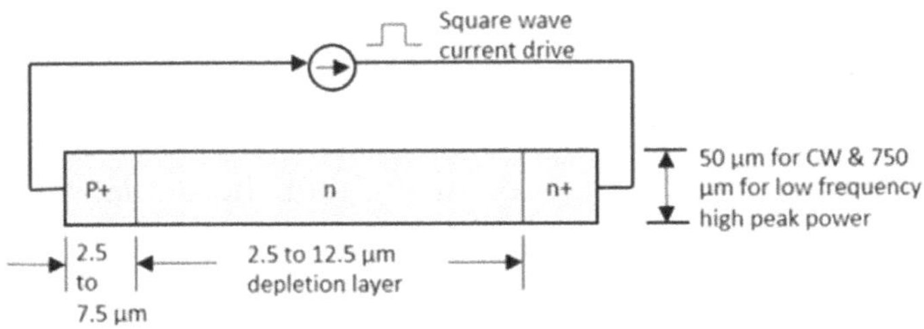

Arrangement in TRAPATT diode

The electrons and holes trapped in low field region behind the zone, are made to fill the depletion region in the diode.

This is done by a high field avalanche region which propagates through the diode.

The following figure shows a graph in which AB shows charging, BC shows plasma formation, DE shows plasma extraction, EF shows residual extraction, and FG shows charging.

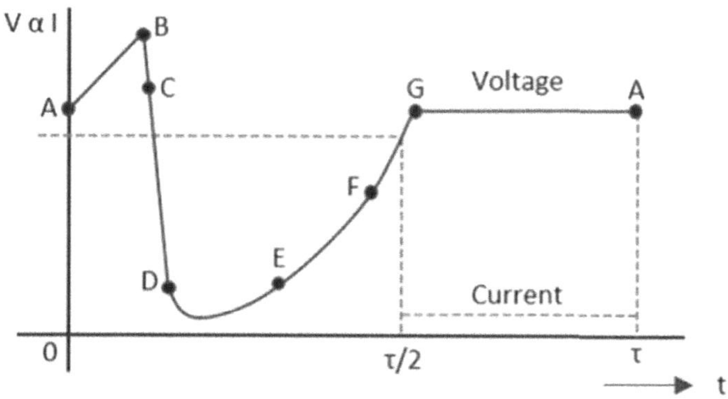

Let us see what happens at each of the points

A: The voltage at point A is not sufficient for the avalanche breakdown to occur. At A, charge carriers due to thermal generation results in charging of the diode like a linear capacitance.

A-B: At this point, the magnitude of the electric field increases. When a sufficient number of carriers are generated, the electric field is depressed throughout the depletion region causing the voltage to decrease from B to C.

C: This charge helps the avalanche to continue and a dense plasma of electrons and holes is created. The field is further depressed so as not to let the electrons or holes out of the depletion layer, and traps the remaining plasma.

D: The voltage decreases at point D. A long time is required to clear the plasma as the total plasma charge is large compared to the charge per unit time in the external current.

E: At point E, the plasma is removed. Residual charges of holes and electrons remain each at one end of the deflection layer.

E to F: The voltage increases as the residual charge is removed.

F: At point F, all the charge generated internally is removed.

F to G: The diode charges like a capacitor.

G: At point G, the diode current comes to zero for half a period. The voltage remains constant as shown in the graph above. This state continues until the current comes back on and the cycle repeats.

The avalanche zone velocity Vs is represented as

$$V_s = \frac{dx}{dt} = \frac{J}{qN_A}$$

Where

- J= Currentdensity
- q= Electron charge 1.6 x10^{-19}
- N_A= Dopingconcentration

The avalanche zone will quickly sweep across most of the diode and the transit time of the carriers is represented as

$$\tau_s = \frac{L}{V_s}$$

Where,

- Vs = Saturated carrier drift velocity
- L = Length of the specimen

The transit time calculated here is the time between the injection and the collection. The repeated action increases the output to make it an amplifier, whereas a microwave low pass filter connected in shunt with the circuit can make it work as an oscillator.

Applications

There are many applications of this diode.

- Low power Doppler radars
- Local oscillator for radars
- Microwave beacon landing system
- Radio altimeter
- Phased array radar, etc.

2.10 Baritt Diode

The full form of BARITT Diode is BARrier Injection Transit Time diode. These are the latest invention in this family. Though these diodes have long drift regions like IMPATT diodes, the carrier injection in BARITT diodes is caused by forward biased junctions, but not from the plasma of an avalanche region as in them.

In IMPATT diodes, the carrier injection is quite noisy due to the impact ionization. In BARITT diodes, to avoid the noise, carrier injection is provided by punch through of the depletion region. The negative resistance in a BARITT diode is obtained on account of the drift of the injected holes to the collector end of the diode, made of p-type material.

The following figure shows the constructional details of a BARITT diode.

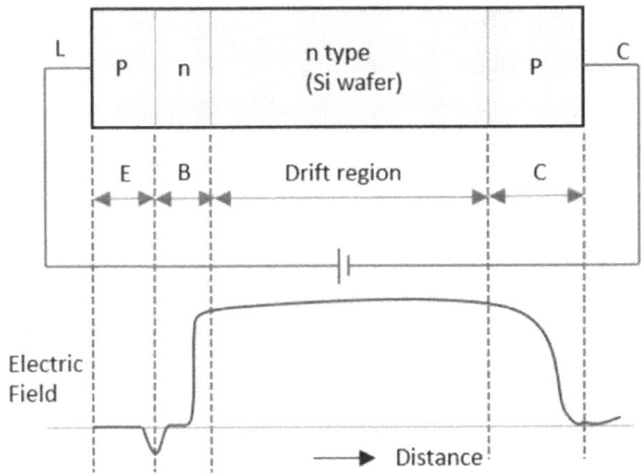

Construction of BARITT diode

For a m-n-m BARITT diode, Ps-Si Schottky barrier contacts metals with n-type Si wafer in between. A rapid increase in current with applied voltage above30v is due to the thermionic hole injection into the semiconductor.

The critical voltage (Vc) depends on the doping constant (N), length of the semiconductor (L) and the semiconductor dielectric permittivity (ϵS) represented as

$$V_c = \frac{qNL^2}{2\epsilon S}$$

Monolithic Microwave Integrated Circuit MMIC

Microwave ICs are the best alternative to conventional waveguide or coaxial circuits, as they are low in weight, small in size, highly reliable and reproducible. The basic materials used for monolithic microwave integrated circuits are –

- Substrate material
- Conductor material

- Dielectric films
- Resistive films

These are so chosen to have ideal characteristics and high efficiency. The substrate on which circuit elements are fabricated is important as the dielectric constant of the material should be high with low dissipation factor, along with other ideal characteristics. The substrate materials used are GaAs, Ferrite/garnet, Aluminum, beryllium, glass and rutile.

The conductor material is so chosen to have high conductivity, low temperature coefficient of resistance, good adhesion to substrate and etching, etc. Aluminum, copper, gold, and silver are mainly used as conductor materials. The dielectric materials and resistive materials are so chosen to have low loss and good stability.

Fabrication Technology

In hybrid integrated circuits, the semiconductor devices and passive circuit elements are formed on a dielectric substrate. The passive circuits are either distributed or lumped elements, or a combination of both.

Hybrid integrated circuits are of two types.

- Hybrid IC
- Miniature Hybrid IC

In both the above processes, Hybrid IC uses the distributed circuit elements that are fabricated on IC using a single layer metallization technique, whereas Miniature hybrid IC uses multi-level elements.

Most analog circuits use meso-isolation technology to isolate active n-type areas used for FETs and diodes. Planar circuits are fabricated by implanting ions into semi-insulating substrate, and to provide isolation the areas are masked off.

"Via hole" technology is used to connect the source with source electrodes connected to the ground, in a GaAs FET, which is shown in the following figure.

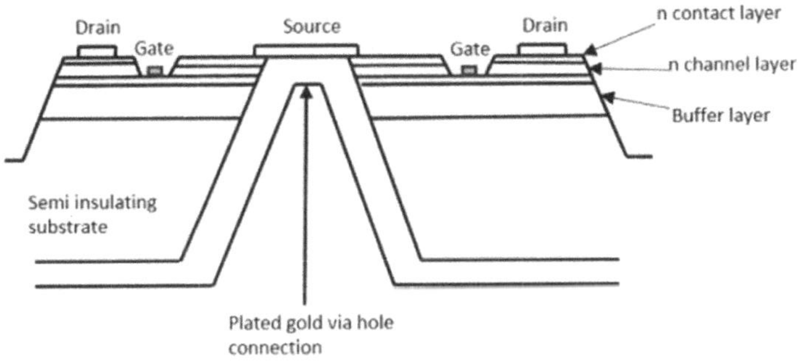

Via-hole technology in monolithic IC's

There are many applications of MMICs.

- Military communication
- Radar
- ECM
- Phased array antenna systems
- Spread spectrum and TDMA systems

They are cost-effective and also used in many domestic consumer applications such as DTH, telecom and instrumentation, etc.

2.11 Read Diode -IMPATT Diodes

The IMPATT diode or IMP actionisation Avalanche Transit Time diode is an RF semiconductor device that is used for generating microwave radio frequency signals.

The IMPATT diode technology is able to generate signals typically from about 3 and 100 GHz or more. One of the main advantages of this microwave diode is the relatively high power capability (often ten watts and more) which is much higher than many other forms of microwave diode.

Although the IMPATT diode is not as widely used these days as other technologies have been able to provide higher levels of performance, it nevertheless fits a niche in the microwave signal generation market, especially where relatively cost effective sources are needed.

IMPATT Discovery & Development

The original idea for IMPATT diode technology was proposed by William Shockley in 1954. He developed the concept for creating a negative resistance using a transit time delay mechanism. In his proposal, the method of injection for the carriers was a forward biased PN junction.

Shockley published his idea in the Bell Systems Technical Journal in 1954 in a paper entitled: 'Negative resistance arising from transit time in semiconductor diodes.'

The idea was not developed further until 1958 when W.T. Read of Bell Laboratories proposed his P+ N I N+ diode structure which was later called the Read diode. This diode used avalanche multiplication as the injection mechanism. Again a paper was published in the Bell Systems Technical Journal, this time in 1958 under the title: A proposed high-frequency, negative resistance diode.'

Although the injection mechanism and diode had been postulated, it was not possible to realise the diode at first. It took until 1965 before the first practical operating diodes were made and the first oscillations were observed. The diode used for this demonstration was fabricated using silicon and had a P+ N structure.

After this, operation of the Read diode was demonstrated and then in 1966 a PIN diode was also demonstrated to work.

IMPATT Diode Basics

In many respects the IMPATT diode is an unusual diode in that it is able to provide high power RF signals at microwave frequencies using a structure that is not that far different from the basic PN junction. However it has been developed to enable a totally different mode of operation to be utilised.

- **Theory & operation:** The IMPATT microwave diode relies upon a negative resistance effect caused by the transit time of the carriers. When a negative resistance occurs, reducing the voltage increases the current and vice versa. This negative resistance enables the diode to act as an oscillator, creating signals at microwave frequencies. Read more about the How does an IMPATT diode work?
- **Fabrication & structure:** There are a number of structures and fabrication methods used for IMPATT diodes. Each one has its own advantages and disadvantages. Read more about the IMPATT structure & fabrication

IMPATT Diode Operation

The IMPATT diode is generally used in microwave generators. To create an output, a DC supply applied to the IMPATT which oscillates when a suitable tuned circuit is in circuit.

The IMPATT circuit output is reliable and relatively high when compared to other forms of microwave diode. However it also generates high levels of phase noise as a result of its operation and this means that it is used in simple transmitters more frequently than as a local oscillator in receivers where the phase noise performance is generally more important. The phase noise issue also means that it has to be used in relatively undemanding applications where the phase noise performance is unlikely to be of importance.

The IMPATT diode requires a relatively high voltage for its operation. Often this may be as high as 70 volts or possibly more. Along with the phase noise, this can limit the applications in which the diode can be used. Nevertheless IMPATT diodes are particularly attractive option for microwave diodes for many areas.

IMPATT Diode Applications

IMPATT diodes are used in a number of applications where a compact cost effective ,means of generating microwave power is required. They have the drawback that on their own they are free running, and also in view of the way in which they operate, they generate relatively high levels of phase noise. Nevertheless there are several applications for which IMPATT diode technology is applicable.

- Intruder alarms
- Basic forms of radar
- General detectors using RF technology

There are many areas in which the IMPATT diode can be used, but with other alternatives now becoming available, along with its high phase noise and the high voltage needed, its use has declined. Nevertheless it is still a form of diode that can be useful in some applications.

2.12 Parametric Devices

These reactance-diode parametric amplifiers are designed for applications at all microwave frequencies. The heart of the line is the gallium- arsenide, variable-reactance diode, which gives improved noise performance over a wide range of...

These reactance-diode parametric amplifiers are designed for applications at all microwave frequencies. The heart of the line is the gallium- arsenide, variable-reactance diode, which gives improved noise performance over a wide range of temperatures. Typical models in the series include the following:

- Model S-21, an S-band amplifier, which is tunable from 2.7 to 2.9 gigacycles, and carries ratings of 3-db over-all noise figure (double sideband) with a 15-db gain over a 15-mc bandwidth;
- Model C-11, a C-band amplifier tunable from 4.4 to 5 gigacycles, with a 3-db over-all noise figure (double sideband) with 15-db gain over a 20-mc bandwidth;
- Model L-11, tunable from 0.9 to 1.2 gigacycles, with a noise figure of 1 db (double sideband) with 16-db gain over a 50-mc bandwidth. For further information, contact Texas Instruments, 6000 Lemmon Ave., Dallas, Tex. (Electronic Design, Oct. 26, 1960, p. 112)
- Texas Instruments made many contributions to rf and microwave semiconductor devices and circuitry, even at the amplifier equipment level, such as these parametric amplifiers. Paramps reduced the noise figure of microwave receivers without the complexity and cost of masers. The tunnel diode was another contender for such applications, but its negative resistance characteristics were difficult to handle, and it fell out of favor among designers.

UNIT - III

MICROWAVE TUBES

Klystrons - Two cavity Klystron Amplifiers - Reflex Klystrons - Velocity Modulation - Power Output and Efficiency - Electronic Admittance - Helix Traveling - Wave Tubes (TWTs) - Slow-Wave structures - Amplification Process - Convection Current - Axial Electric Field - Wave Modes - Gain Consideration - Magnetron Oscillators - Cylindrical Magnetron - Coaxial Magnetron.

3.1 Klystrons

What is a Klystron Tube?

A Klystron (also known as a Klystron Tube or Klystron Amplifier) is a vacuum tube that is used to oscillate and amplify microwave frequency signals. It was invented by American electrical engineers Russell and Sigurd Varian.

A klystron uses the kinetic energy of an electron beam. Generally, Low-power klystrons are used as oscillators and high-power klystrons are used as output tubes in UHF.

There are two configurations for a low-powered klystron. One is a low-power microwave oscillator (Reflex Klystron) and the second is a low-power microwave amplifier (Two Cavity Klystron or Multi Cavity Klystron).

What is a Reflex Klystron Oscillator?

Before answering this question, we need to know how the oscillations are generated. To generate the oscillations, we need to give positive feedback from the output to input. With constraint that the loop gain is unity.

For a klystron, the oscillations will generate if a part of the output is used as feedback to the input cavity and keep the loop gain magnitude unity. The phase shift of the feedback path is one cycle (2π) or multiple cycles (multiple of 2π).

Construction of Reflex Klystron

The electron beam is injected from the cathode. Then there is an anode, known as focusing anode or accelerating anode. This anode is used to narrow down the electron beam. The anode is connected with the positive polarity of the DC voltage source.

The reflex klystron has only one cavity, which is placed next to the anode. This cavity works as a buncher cavity for forward-moving electrons and catcher cavity for backward moving electrons.

The velocity and current modulation take place in the cavity gap. The gap is equal to the distance ‗d'.

The repeller plate is connected with the negative polarity of the voltage source Vr.

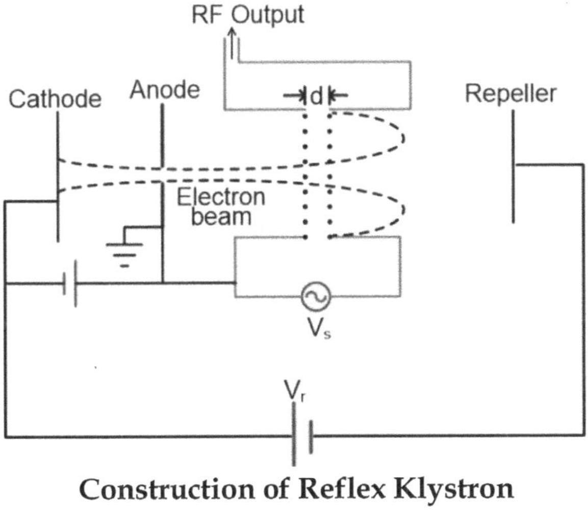

Construction of Reflex Klystron

Working Principle of Reflex Klystron

Reflex Klystron works on the principle of velocity and current modulation.

The electron beam is injected from the cathode. The electron beam passes through the accelerating anode. The electron moves in the tube with uniform velocity until it reaches the cavity.

The velocity of electrons is modulated in the cavity gap and these electrons try to reach the repeller.

The repeller is connected with the negative polarity of a voltage source. Hence, because of the same polarity, it opposes the force of electrons.

The kinetic energy of electrons decreases in the repeller space and at some point, it will be zero. After that, the electron pulls back to the cavity. And in the return journey, all electrons bunched at the one point.

There will be current modulation due to the bunch formation. The energy of electrons are converted in the form of RF and RF output is taken from the cavity. For maximum efficiency of the klystron, the bunching of the electron must take place in the center of the cavity gap.

How do electrons moves in the klystron tube?

From the electron gun (cathode), the electron beam is injected in the tube. These electrons move towards the anode with uniform velocity. Then electrons pass through the cavity gap. The velocity of electrons varies according to the cavity gap voltage.

If the cavity gap voltage is positive, the electron will be accelerated and if the cavity gap voltage is negative, the electron will be decelerated. If the voltage is zero, the velocity of electrons will not change.

When electrons leave from the cavity gap, all electrons have different velocities and these electrons will travel in the repeller space.

These electrons travel the distance according to the velocity. Higher the velocity, the electron will travel more distance and lower the velocity, the electron will travel less distance in the repeller space.

All these electrons will return to the cavity and bunched at the center of the cavity gap. The energy of electrons transferred from the cavity is known as the RF output.

Apple-gate Diagram

Apple-gate diagram is a graph between the distance from the cavity gap and the time taken by the electron in the repeller space.

Different electrons follow different paths depending upon their velocities. The velocity of electrons depends on the cavity gap voltage.

Let's take the example of three electrons. The reference electron (e0) enters the cavity gap when the cavity gap voltage is zero. Hence, the velocity will not change. It travels L0 distance in the repeller space and pulls back to the cavity. Because of the repeller plate is highly negative and it will oppose the kinetic energy of an electron.

The electron enters before e0, this electron is known as an early electron (ee). This electron enters the cavity gap when the cavity gap voltage is positive. Therefore, the electron velocity will increase. It will travel Le distance and pull back to the cavity.

The electron enters after e0, this electron is known as a late electron (el). This electron enters the cavity gap when the cavity gap voltage is negative. Therefore, the electron velocity will decrease. It will travel Ll distance and pull back to the cavity.

The graph below helps explain this process:

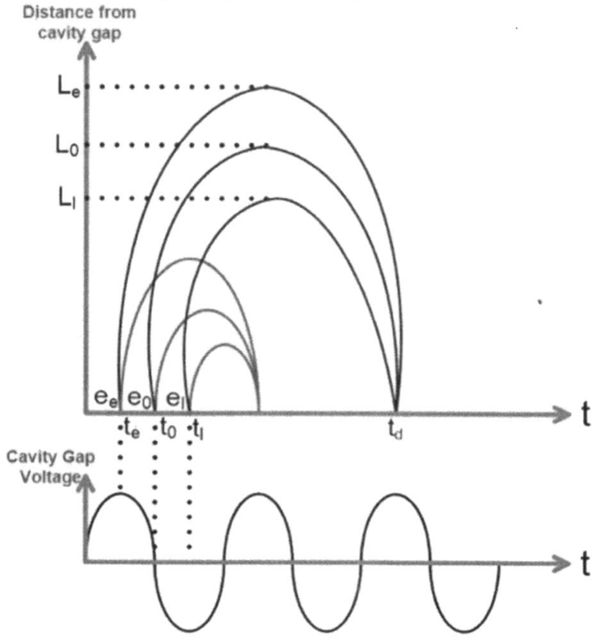

Bunching Process

The total time needs from the cavity gap to repeller space and repeller space to cavity gap are the same for all electrons. And that time is Td.

At time Td all electrons bunched at the center of cavity gap. And this is how the bunching process occurs in the klystron tube.

Applications of Reflex Klystron

The applications of a Reflex Klystron include:

- Radio and RADAR receiver
- A signal source in microwave generators

- Frequency modulated oscillator in portable microwave links
- Pump oscillator for parametric amplifiers
- Local oscillator in microwave receivers

3.2 Two Cavity Klystron

The working principle of Two Cavity Klystron is the same as the Reflex Klystron. The construction diagram of two-cavity klystron is as shown in the below figure.

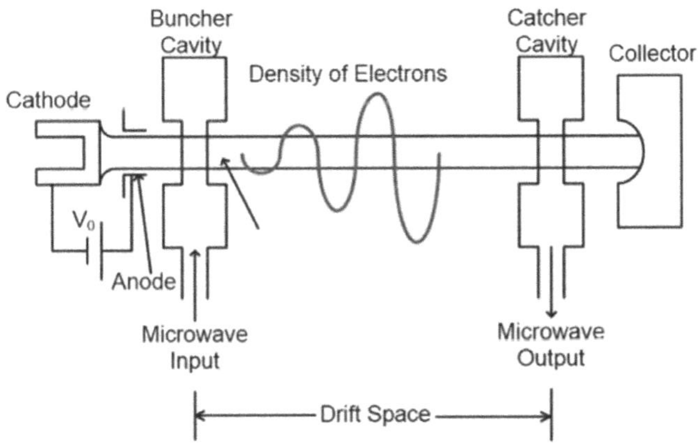

Construction of Two-cavity Klystron

As the name suggests, there are two cavities; the first cavity is a buncher cavity or input cavity and another cavity is the catcher cavity or output cavity. Electrons injected from the cathode and reach with uniform velocity till the buncher cavity.

The input RF signal is given at the input cavity and output signal collected from the output cavity. There is a gap in both cavities, these gaps are known as microwave interaction region.

In the first cavity, the velocity of electrons modulated by the input RF signal present in the first cavity. This called velocity modulation.

It formulates the bunching of electrons. And passing through the catcher cavity. The current modulation takes place in the catcher cavity.

After passing the second cavity, all electrons lose their kinetic energy to the microwave fields. They will be collected by collectors because of the reduction of velocity.

Klystron vs Magnetron

The Klystron is a vacuum tube which used as an oscillator and amplifier of microwave signals. The magnetron is different from the klystron tube. The magnetron used only as an oscillator.

In a klystron, the electron is injected normally from the cathode. But in the case of magnetron, the electrons are forcefully injected.

In a klystron, the electrons move linearly in the tube and in the magnetron, the electrons follow a spiral path from cathode to anode.

The klystron is used in TV transmitters, RADARs and particle accelerators. It is also used as a high power, narrowband stable amplifier. The magnetron used in microwave ovens, operating at 2.45 GHz. It is also used for RF heating, operating at 900 MHz or 2.45 GHz.

Klystron Specifications

Frequency of operation: 1 to 200 GHz
Bandwidth: +/- 30 MHz
Power Output: 10 mW to 2.5 W
Practical Efficiency: 10-20%
Theoretical Efficiency: 22.78%
Tuning Range: 5 GHz @ 2W, 30 GHz @ 10mW

Magnetron Specification

Frequency Range: 500 MHz to 12 GHz
Power: 600W @ 2.45 GHz
Peak Power: 40MW with DC voltage of 50 kV @10 GHz
Average Power: 800kW
Duty Cycle: 0.1%
Efficiency: 40 to 70 %

Difference Between Two Cavity Klystron and Reflex Klystron

In two-cavity klystron, the buncher cavity and catcher cavity is different. But in reflex klystron, there is only one cavity. And this cavity works as a buncher cavity as well as a catcher cavity.

In two-cavity klystron, the collector used to collect the electron. Whereas in the reflex klystron, the repeller plate used in place of collector and used to repel the electron beam.

The main purpose of Two cavity klystron is to amplify the microwave signal. The main purpose of the reflex klystron is to oscillate the microwave signal.

Klystron - Two-cavity Klystron Amplifier

Definition: Klystrons are a special type of vacuum tubes that find applications as amplifiers and oscillators at microwave frequencies. Its principle of operation is velocity modulation. Thus the device used for amplifying microwave signals is known as Two-cavity Klystron.

In the year 1937, American engineers Russell and Sigurd Varian developed klystrons. Basically, the operating principle of Klystron is such that the kinetic energy of a moving electron beam is utilized for amplifying and generating microwave signals. TWT i.e., Travelling Wave Tubes are also used for amplification of RF signals and have similar applications as klystrons. But in TWT, a continuous interaction is maintained between the field and electron beam. While in klystrons interaction between the two is allowed to occur only at the cavities of the structure.

Klystrons are majorly classified as:

- Two-cavity Klystron
- Reflex Klystron

Here in this article, we will discuss the construction and working of the two-cavity klystron amplifier.

Two-cavity Klystron

- Operating Principle
- Construction
- Working
- Applications

Operating Principle of Two-cavity Klystron

As we have already discussed in the introduction that Klystron is based on the principle of velocity modulation. Thus two-cavity klystron amplifier utilizes the kinetic energy of moving electron beam for signal amplification.

The variation in the velocity of electrons while moving inside the tube is known as velocity modulation. This velocity modulation permits bunching of electrons while propagation. So, the combined energy of bunched electrons is transferred at the output thereby providing an amplified signal.

Construction

The figure below represents the structure of a two-cavity klystron:

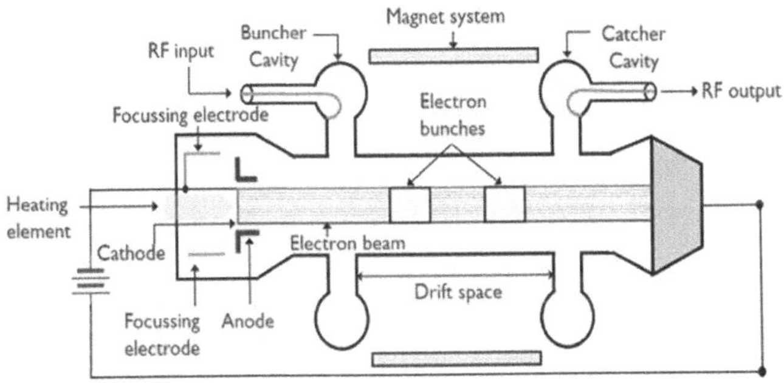

As we can see that the above figure consists of 2 cavities namely the buncher cavity and catcher cavity. The RF signal to be amplified is provided at the buncher cavity. The electron gun comprises cathode, heating element and anode. The electron beam is produced by the cathode by making use of a heating element and the high positive potential at the anode provides the required acceleration to the electron beam initially. The region between two cavities is known as drift space.

To allow focussed propagation of electron beam inside the tube an external electromagnetic winding is used that generates a longitudinal magnetic field. This is done in order to prevent the spreading of the beam inside the tube. The amplified RF signal is achieved at the catcher cavity. Also, a collector is present near the second cavity that collects the electron bunch.

Working of Two-cavity Klystron Amplifier

Consider the above-shown figure of two-cavity klystron amplifier.

Initially, electrons are emitted from the electron gun and the anode present in the structure provides the desired acceleration to the beam.

- In the absence of any RF input, the electron will tend to move with their respective uniform velocities to reach the catcher cavity and gets collected at the collector.
- But when external RF signal is applied at the input of the buncher cavity then this causes the generation of a local electric field inside the tube.

This electric field causes the bunching of electrons as the field applies acceleration and deceleration to the moving electron, according to the polarity of the signal by which the field is generated.

Basically, the reason for causing acceleration and deceleration is that when the direction of movement of an electron is opposite to the direction of the field, then, in this case, the electrons experience a decrease in their moving velocity. However, if the generated electric field and the direction of movement of the electron are the same then, in this case, the electrons experience increase in the velocity of their movement.

Let us now understand in detail how this increase and decrease in velocity causes bunching of electrons:

- When the negative half of the RF signal is provided as input to the buncher cavity then the moving electrons experience a repulsive force due to the presence of a negative charge at the entering plate of the buncher cavity.deceleration in two cavity klystron

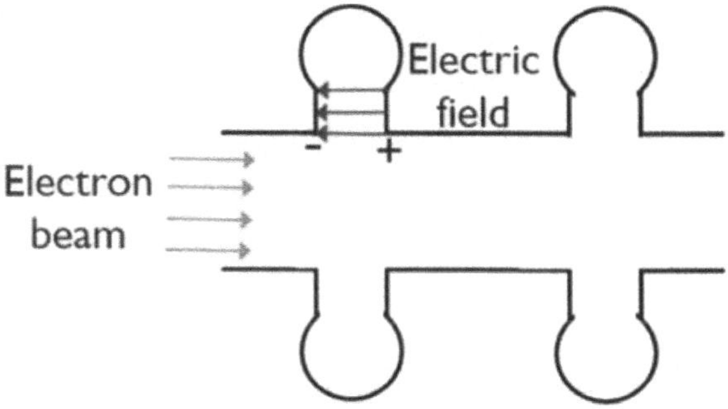

Or we can say, that due to the negative half of the input the generated field will be in a direction opposite to the direction of the movement of electrons. So, because of the opposition offered by the field, the moving velocity of electrons gets reduced.

Further when the positive half of the RF signal is provided then the positive potential at the first plate of the cavity applies attractive force to the moving electrons. More simply, for the positive half cycle of input, the generated electric field will be in a direction similar to the direction of electron movement.acceleration in two cavity klystron.

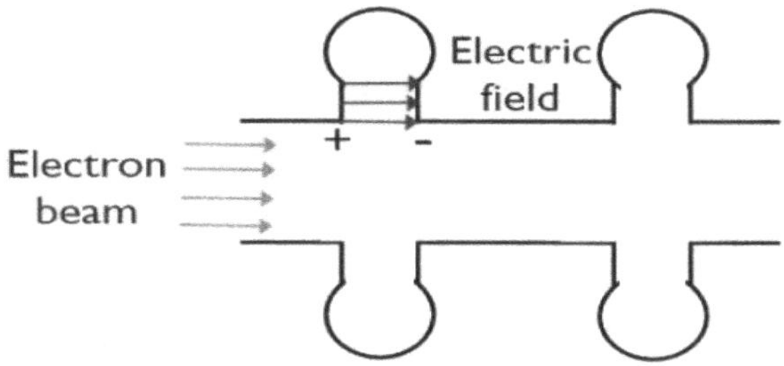

So, this leads to an increase in the moving velocity of the electrons.

Thus, combinely when we consider both the cases then the electrons that were emitted earlier by the gun will be decelerated. While the electrons emitted later will be accelerated. Thus all the electrons while moving with different velocities get bunched in the drift space. This change in the velocity of electrons while moving due to RF input is known as velocity modulation. Once the electron bunching is done then the catcher cavity present at another end of the tube absorbs the beam energy.

It is to be noted that to find the position of the catcher cavity transit time of the bunches must be considered. This is so because the catcher cavity must be present at a sufficient distance from the buncher cavity so that bunching can be attained in the drift space. Further, once the energy is transferred to the catcher cavity then electrons (now with low energy) gets collected at the collector.

Applegate Diagram

The figure below shows the Applegate diagram that represents the bunching of electrons moving with different velocities:

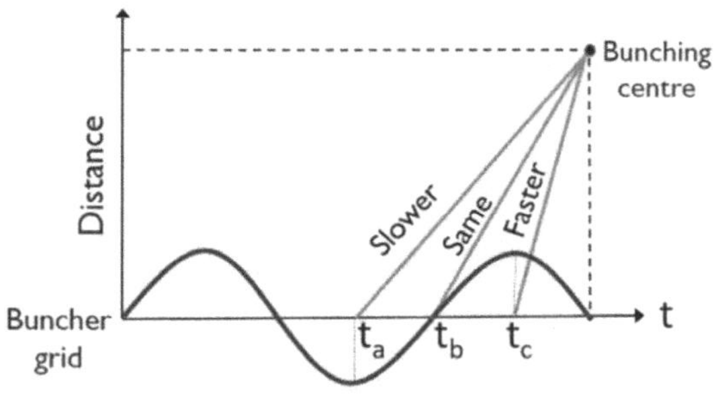

The electron travelling inside the tube under the absence of external fields acts as the bunching centre. Also, the electrons moving due to the influence of the positive half cycle of the signal reaches faster. While the movement due to the negative half cycle is retarded. Thus the figure represents the bunching process at a certain point and at a specific distance inside the tube.

Applications

The two-cavity Klystron finds application in satellite communication, UHF TV transmitters as well as radar systems, wideband high power communication and troposphere scatter transmitters etc.

3.3 Reflex Klystrons

This microwave generator, is a Klystron that works on reflections and oscillations in a single cavity, which has a variable frequency.

Reflex Klystron consists of an electron gun, a cathode filament, an anode cavity, and an electrode at the cathode potential. It provides low power and has low efficiency.

Construction of Reflex Klystron

The electron gun emits the electron beam, which passes through the gap in the anode cavity. These electrons travel towards the Repeller electrode, which is at high negative potential. Due to the high negative field, the electrons repel back to the anode cavity. In their return journey, the electrons give more energy to the gap and these oscillations are sustained. The constructional details of this reflex klystron is as shown in the following figure.

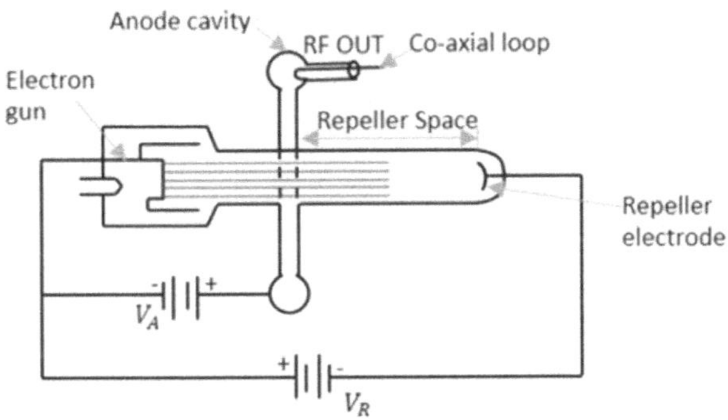

Constructional details of Reflex Klystron

Reflex Klystron

It is assumed that oscillations already exist in the tube and they are sustained by its operation. The electrons while passing through the anode cavity, gain some velocity.

Operation of Reflex Klystron

The operation of Reflex Klystron is understood by some assumptions. The electron beam is accelerated towards the anode cavity.

Let us assume that a reference electron er crosses the anode cavity but has no extra velocity and it repels back after reaching the Repeller electrode, with the same velocity. Another electron, let's say ee which has started earlier than this reference electron, reaches the Repeller first, but returns slowly, reaching at the same time as the reference electron.

We have another electron, the late electron e_l, which starts later than both e_r and e_e, however, it moves with greater velocity while returning back, reaching at the same time as er and e_e.

Now, these three electrons, namely e_r, e_e and e_l reach the gap at the same time, forming an electron bunch. This travel time is called as transit time, which should have an optimum value. The following figure illustrates this.

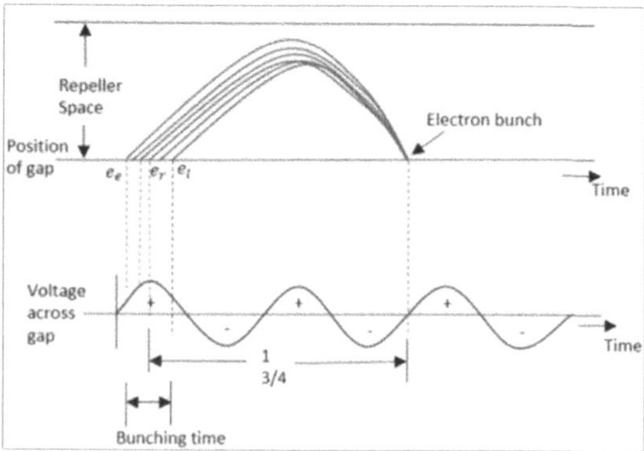

The anode cavity accelerates the electrons while going and gains their energy by retarding them during the return journey. When the gap voltage is at maximum positive, this lets the maximum negative electrons to retard.

The optimum transit time is represented as

T = n+34wherenisaninteger

This transit time depends upon the Repeller and anode voltages.

Applications of Reflex Klystron

Reflex Klystron is used in applications where variable frequency is desirable, such as −

- Radio receivers
- Portable microwave links
- Parametric amplifiers
- Local oscillators of microwave receivers
- As a signal source where variable frequency is desirable in microwave generators.

3.4 Velocity Modulation

Klystron Amplifiers

The 2 cavity klystron amplifier is widely used microwave amplifier, operated by the principles of voltage and current modulators.

Basic Operations

A high velocity electron beam produced by the accelerating anode is passed through a buncher cavity, drift space, catcher cavity and finally collected by the collector terminals. The electrons injected from the cathode is accelerated by applying a DC voltage ‗Vg'. They arrive at the first cavity that is the buncher cavity or input cavity with uniform velocity. At the buncher cavity these electrons encounter signal voltage or gap voltage. The electrons that pass through the zeros of the gap voltage pass with unchanged velocity. The electrons that pass through positive half cycles of the gap voltage undergo acceleration in velocity. The electrons that pass through negative half cycles of the gap voltage undergo retardation in velocity. (As a result of these the electrons get bunched together as they travelled through the drift space). The variation in electron velocity in drift space is called velocity modulation. (The buncher cavity velocity modulates the electron beam). This electron beam induces a RF current in this field is opposite to the input cavity. Thus the kinetic energy is transferred from the electrons to the field capture cavity. The second cavity is called capture cavity since it captures energy from the bunch electron beam. The electrons emerging from the capture cavity are collected by the collector terminal.

Velocity Modulation in Klystron Amplifier

The velocity of electrons before entering the buncher cavity is given by,

$Vo = \sqrt{2ev0/m}$

Where m is the mass of the electron

e is the charge of electron

v0 is the cathode potential

On substituting the values of e and m, the equation reduces to

$Vo = 0.596 \times 106 \sqrt{(v0)}$ m/sec ------------ 1

When the microwave signal is applied to the input terminal, the gap voltage is given by

$Vs = V1 \sin\omega t$ --------------- 2

V1 is the amplitude of the signal.

The average transit time through the gap at a distance ‗d‘.

$T = d/Vo = t1 - t0$ ------------- 3

Where, t0 is the line at which beam reaches the buncher cavity. t1 is the time at which the beam leaves the buncher cavity.

The average transit angle,

$\theta g = \omega(d/Vo) = \omega(t1 - t0)$ ----------- 4

The average microwave voltage in the buncher cavity is

$$< Vs >= \frac{1}{T}\int_{t1}^{t0} V_1 sinwtdt$$

$= V1/T\,\omega\,(\cos \omega t0 - \cos \omega t1)$ ---------------- 5

From eq (4)

$\omega d/V0 = \omega(t1 - t0)$

$\omega t1 = \omega(d/Vo + t0)$ ------------ 6

Subsitutingeq (6) in eq (5)

$<Vs> = V1/T\omega\,[\cos\omega t0 - \cos(\omega d/Vo + \omega t0)]$ ------------ 7

Let $\omega t0 + \omega d/2Vo = \omega t0 + \theta g/2 = A$

$\omega d/2Vo = \theta g/2 = B$

$A + B = \omega t0 + \theta g/2 + \theta g/2 = \omega t0 + \theta g$

$A - B = \omega t0 + \theta g/2 - \theta g/2 = \omega t0$

$A + B = \omega t0 + \theta g\ ,\ A - B = \omega t0$ ---------------- 8

Substitute eq (8) in eq (7)

$<Vs> = V1/T\omega\,[\cos(A - B) - \cos(A + B)]$

$= 2V1/T\omega.\ \sin A \sin B$

$= 2V1/T\omega.\ \sin(\omega t0 + \theta g)\sin(\theta g/2)$

Let $\omega T = \theta g$

Therefore, $<Vs> = 2V1/\theta g.\ \sin(\omega t0 + \theta g)\sin(\theta g/2)$

$<Vs>= V1\,\beta i\,\sin(\omega t0 + \theta g/2)$

$\boldsymbol{\beta i = \sin(\theta g/2)/(\theta g/2)}$

3.5 Power Output and Efficiency

The equivalent output circuit of Klystron is

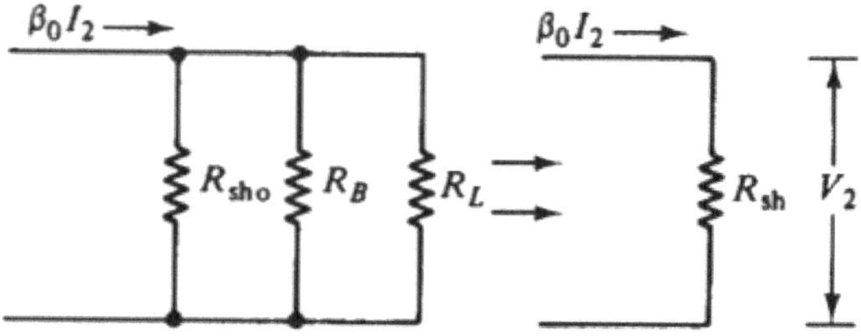

Where R_{sho} = Wall resistance of catcher cavity

R_B = beam loading resistance

R_L = External load resistance

R_{sh} = Effective shunt resistance = $R_{sho} \| R_B \| R_L$

$$i_{rms} = \frac{i_{f\,max}}{\sqrt{2}} = \frac{I_2}{\sqrt{2}} = \frac{2I_0\beta_0 J_1(X)}{\sqrt{2}}$$

$$Poutput = i^2_{rms}\ Rsh = 2I_0^2\beta_0^2 J_1^2(X)R_{sh} = \beta_0 I_0 J_1(X)V_2$$

$$\max imum\ output\ voltage = V_2 = i_{f_{rms}}\ Rsh = 2\beta_0 I_0 J_1(X)R_{sh}$$

$$Efficiency = \eta = \frac{P_0}{P_{dc}} = \frac{2I_0^2\beta_0^2 J_1^2(X)R_{sh}}{V_0 I_0}$$

$$\eta = \frac{\beta_0 I_0 J_1(X)V_2}{V_0 I_0} = \beta_0 J_1(X)\frac{V_2}{V_0}$$

Maximum theoretical efficiency of Klystron is

$\beta_0 = 1$, $J_1(X) = 0.582$, $V_2 = V_0$

η_{max} = 58.2 % theoretical

Practically $\eta \approx 40\%$

Condition for maximum transfer of energy to catcher cavity

3.6 Electronic Admittance

We have seen earlier that the fundamental component of current in the cavity is

$$I_f = 2I_0\beta_i J_1(x^1)\cos(\omega t_2 - \theta_0)$$

The above equation in the phasor form is

$$I_f = 2I_0\beta_i J_1(x^1)e^{-j\theta_0}$$

There is a phase difference of π/2 between the voltages V_1 at $t = t_0$ and V_2 at $t = t_2$

$$V_2 = V_1^{-ej\pi/2}$$

The electronic admittance Y_e is defined as the ratio of i_f to V_2

$$Ye = i_f / V_2$$

$$Y_e = \frac{I_0}{V_0}\beta_i^2\theta'_0 . \frac{J_1(x^1)}{X^1} e^{j\left(\frac{\pi}{2}-\theta_0\right)}$$

where $Y_0 = \frac{I_0}{V_0}$ =dc beam admittance

$$Y_e = Y_0 \frac{\beta_i^2\theta'_0}{} . \frac{J_1(x^1)}{X^1} e^{j\left(\frac{\pi}{2}-\theta_0\right)}$$

Ye is a complex quantity and can be written as $Y_e = G_e + j B_e$

It can be seen that the amplitude of electronic admittance is proportional to dc beam admittance and also transit angle or N

The equivalent circuit of Reflex Klystron is given by

$$Y_e = \frac{I_0}{V_0}\beta_i^2\theta'_0 . \frac{J_1(x^1)}{X^1} e^{j\left(\frac{\pi}{2}-\theta_0\right)}$$

where $Y_0 = \frac{I_0}{V_0}$ =dc beam admittance

$$Y_e = Y_0 \frac{\beta_i^2\theta'_0}{} . \frac{J_1(x^1)}{X^1} e^{j\left(\frac{\pi}{2}-\theta_0\right)}$$

Ye is a complex quantity and can be written as $Y_e = G_e + j B_e$

It can be seen that the amplitude of electronic admittance is proportional to dc beam admittance and also transit angle or N

The equivalent circuit of Reflex Klystron is given by

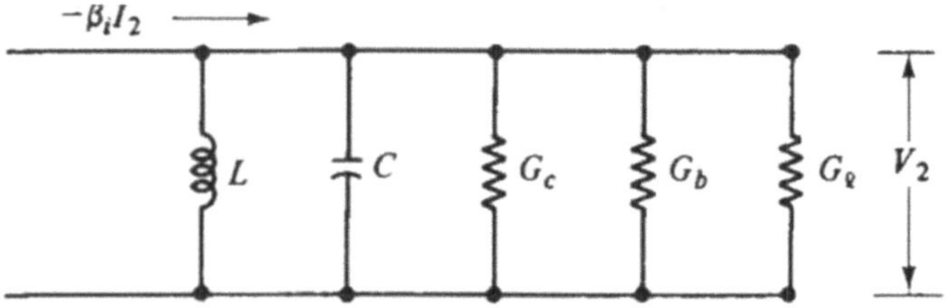

Fig 5.9: Equivalent circuit of reflex klystron

Where G_c = copper loss conductance

G_b = Beam loading conductance

G_L = Load conductance

L and C are energy storage elements of the cavity

G = Conductance of the cavity including external load =

$G = G_c + G_b + G_L = 1 / R_{sh}$

Equation 5.69 can also represented as

$Y_e = G_e + jB_e$

Where G_e = Electronic conductance

B_e = Electronic Sustenance

3.7 Helix Traveling - Wave Tubes (TWTS)

In the previous unit klystrons and reflex klystrons were analysed in some detail. When it comes to study of TWTs it is appropriate to compare the basic operating principles of both TWT and the klystron. In the case of TWT, the microwave circuit is non-resonant and the wave propagates with the same speed as the electrons in the beam. The initial effect on the beam is a small amount of velocity modulation caused by the weak electric fields associated with the traveling wave. Just as in the klystron, this velocity modulation later translates to current modulation, which then induces an RF current in the circuit, causing amplification. However, there are some major differences between the TWT and theklystrons.

- The interaction of electron beam and RF field in the TWT is continuous over the entire length of the circuit, but the interaction in the klystron occurs only at the gaps of a few resonantcavities.
- The wave in the TWT is a propagating wave; the wave in the klystron isnot.
- In the couple cavity TWT there is a coupling effect between the cavities, whereas each cavity in the klystron operatesindependently.
- A traveling-wave tube (TWT) is a Microwave Amplifier with following characteristics:
- Low Power Amplifier: up to 10kW
- Frequency Range: 3 G Hz – 50 GHz
- Wide Band width: about 800MHz
- Power gain: upto 60dB
- Efficiency: 20 – 40 %

Comparison between TWT and Klystron Amplifier

S.No.	Klystron Amplifier	TWT Amplifier
1	Linear beam „O" Type	Linear beam „O" Type
2	Uses input and output resonant cavities	Uses non-resonant wave circuit
3	Narrow band amplifier	Wide band amplifier BW 800 MHz
4	Interaction between electrons and the field is very short	Longer interaction
5	Non propagating wave	Propagating wave

The TWT operates on the principle of slow wave. It is a not resonant „O"-Type microwave device. Its operation is base on the interaction between the waves in the travelling wave structure and the electronic beam. The main elements of the TWT Amplifier are:

- Electron gun
- RFinput
- Magnets
- Attenuator
- Helixcoil
- RFoutput
- Vacuumtube
- Collector

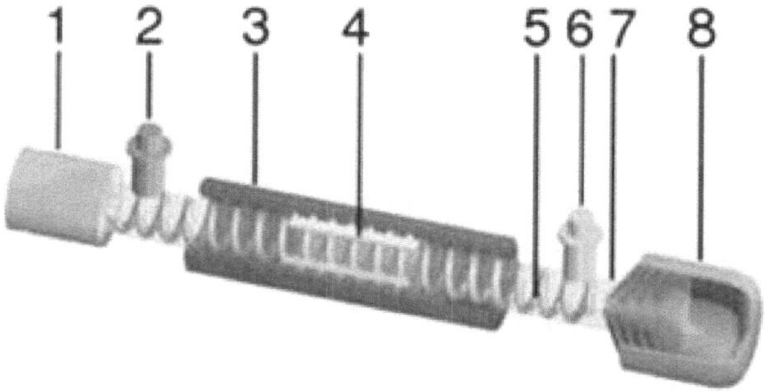

Cutaway view of a TWT

Description

The device is an elongated vacuum tube with an electron gun (a heated cathode that emits electrons) at one end. A magnetic containment field around the tube focuses the electrons into a beam, which then passes down the middle of a wire helix that stretches from the RF input to the RF output, the electron beam finally striking a collector at the other end. The applied RF signal propagates around the turns of the helix and produces and electric field at the center of the helix, with direction of propagation along helix axis.

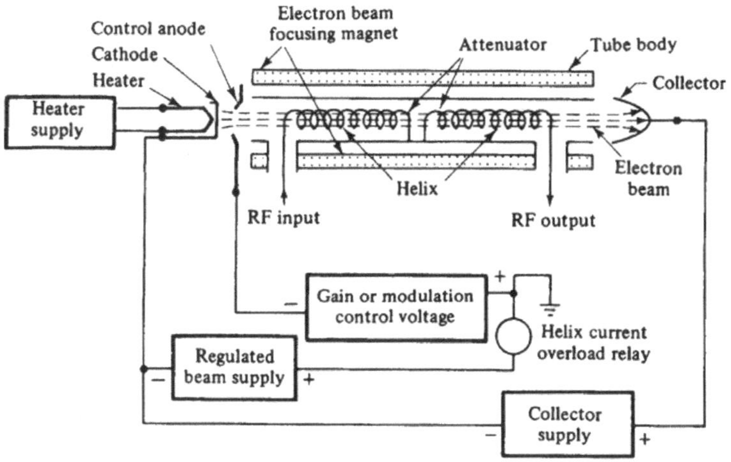

(a)

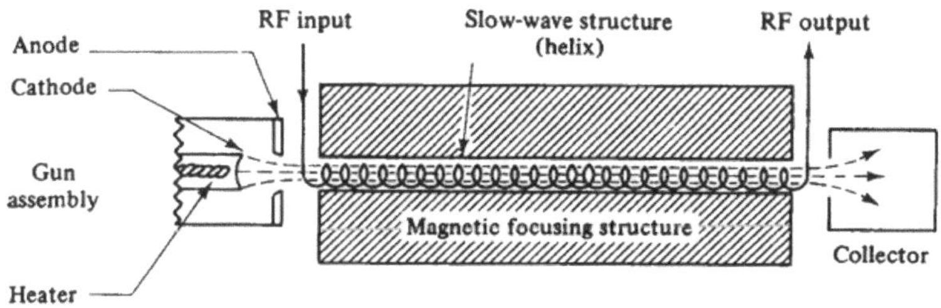

(b)

Helix Traveling Wave Tube (a) Schematic Diagram (b) Simplified Circuit

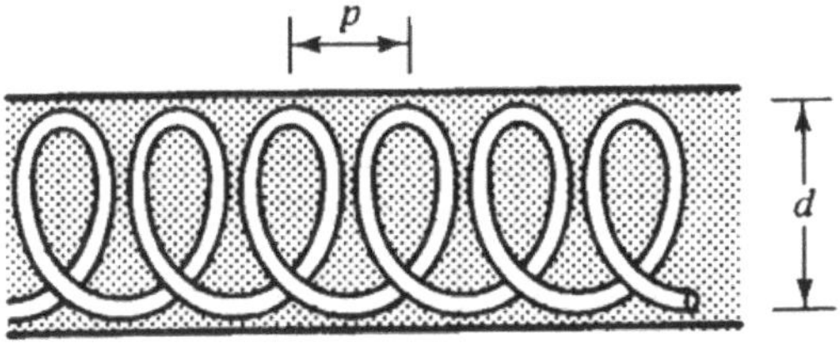

(a)

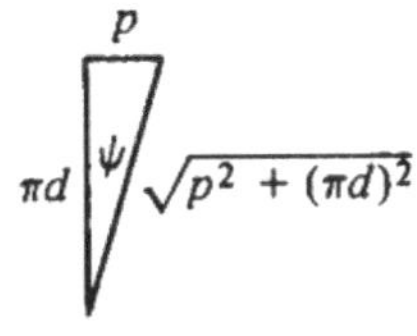

(b)

Helical Slow Wave Structure (a) Helical Coil (b) One turn of helix

3.8 Wave Tubes

This is termed as O-type traveling wave tube. The slow-wave structure is either the helical type or folded back type. The applied signal propagates around the turns of the helix and produces an electric field at the center of the helix, directed along the helix axis. The helix acts as a delay line, in which the RF signal travels at near the same speed along the tube as the electron beam. The axial electric field progresses with a velocity that is very close to the velocity of light multiplied by the ratio of helix pitch to helix circumference. When electrons enter the helix tube, an interaction takes place between the moving axial field and the moving electrons. On the average, the electrons transfer energy to the wave on the helix. This interaction causes the signal wave on the helix to become larger. The electrons entering the helix at zero fields are not affected by the signal wave; those electrons entering the helix at the accelerating field are accelerated, and those at the retarding field are decelerated. As the electrons travel further along the helix, they bunch at the collector end. The bunching shifts the phase by $\pi/2$. Each electron in the bunch encounters a stronger retarding field. Then the microwave energy of the electrons is delivered by the electrons bunch to the wave on helix. The amplification of the signal wave isaccomplished.

An attenuator placed on the helix, usually between the input and output helices, prevents reflected wave from traveling back to the cathode and there by suppresses the oscillations if any.

Higher powered TWT"s usually contain beryllium oxideceramic as both a helix support rod and in some cases, as an electron collector for the TWT because of its special electrical, mechanical, and thermal properties.

3.9 Slow Wave Structures

Slow-wave structures are special circuits that are used in microwave tubes to reduce the wave velocity in a certain directions so that the electron beam and the signal wave can interact. The phase velocity of a wave in ordinary waveguides is greater that the velocity of light in a vacuum. In the operation of traveling wave and magnetron type devices, the electron beam must keep in step with the microwave signal. Since the electron beam can be accelerated only to velocities that are about a fraction of the velocity of light, a slow wave structure must be incorporated in the microwave devices so that the phase velocity of the microwave signal can keep pace with that of the electron beam for effective interactions. Several types of slow-wave structures are shown in the figure givenbelow.

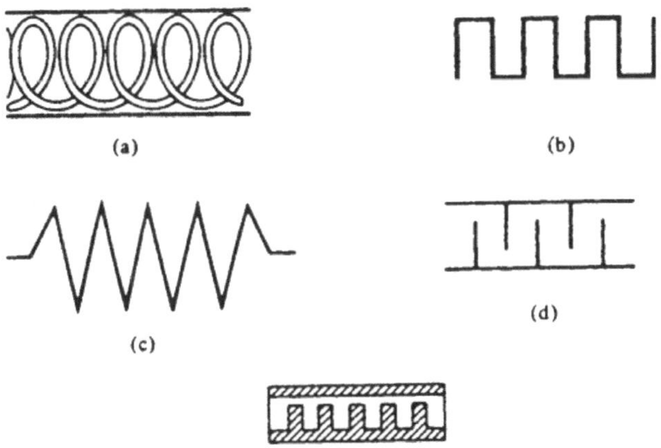

Slow wave structures (a) Helical line (b) Folded-back line (c) Zigzag line (d)Inter-digitalline (e) Corrugatedwaveguide.

3.10 Traveling-Wave Tube Amplifier

- v_P = phase velocity of the em wave along the axial direction = ω / β p = helix pitch

- d = diameter of the pitch ψ = pitch angle
- β = phase constant along the axis. ω = angular frequency of the wave. v_g = Group velocity of the em wave.
- β_o = phase constant of dc electron beam along the axis.
- β_e= phase constant of the velocity modulated electron beam along the axis. v_o = velocity of dc electron beam = $[2eV_o/m]^{1/2}$
- l = length of the slow wave structure.

$$\mathbf{N = l / \lambda_e}$$

Applications

- TWTAs are commonly used as amplifiers in satellitetransponders, where the input signal is very weak and the output needs to be highpower.
- TWT is used as transmitter amplifier particularly in airborne and shop borne fire- control radar systems, Satellites, and in electronic warfare and self-protection systems. In these types of applications, a control grid is typically introduced between the TWT's electron gun and slow-wave structure to allow pulsed operation. The circuit that drives the control grid is usually referred to as a grid modulator.

3.11 Convection Current

Convection Equation Derivation

When the e beam passes through the electric current, it induces a current. This current is known as Convection Current.

The electron velocity, charge density, current density and axial electric field are given by the equations.

v = v0 + v1 e(jωt-γz) ------------------ (a)

$\rho = 3.11\ \rho 0 + \rho 1\ e(j\omega t - \gamma z)$ ------------------

- (b) $J = -J0 + J1\ e(j\omega t - \gamma z)$ -------------------

(c) $E\gamma = -E1e(j\omega t - \gamma z)$ ------------------- (d)

For a small signal, the current density of electron is given by,

$J = \rho v$

Substitute for ρ and v from eqn (a) and (b)

$J = (\rho 0 + \rho 1\ e(j\omega t - \gamma z))(v0 + v1\ e(j\omega t - \gamma z))$

Since the signal is very small, the term $e(j\omega t - \gamma z)$ is neglected.

Therefore, $J = (\rho 0 + \rho 1)(v0 + v1)$

$J = \rho 0\ v0 + \rho 0\ v1 + \rho 1 v0 + \rho 1\ v1$

Let $\rho 1\ v1 = 0$

Therefore, $J1 = \rho 0\ v0 + \rho 0\ v1 + \rho 1 v0$

Substitute for J from eqn (c), the above equation becomes,

$-J0 + J1\ e(jwt - rz) = \rho 0\ v0 + \rho 0\ v1 + \rho 1 v0$

Let, $\rho 0\ v0 = -J0$ and

$J1 = \rho 0\ v1 + \rho 1 v0$ --------------------- (e)

The axial electric field affects the velocity of electrons as,

$dv/dt = -e/mE1\ e(jwt - rz)$ ----------- (1)

Substitute for v from eqn (a)

$dv/dt = \partial/\partial t(v0 + v1\ e(j\omega t - \gamma z)) + \partial z/\partial t.\ \partial/\partial z(v0 + v1\ e(j\omega t - \gamma z))$

$dv/dt = v1\ e(j\omega t - \gamma z)(j\omega - \gamma\ \partial z/\partial t)$

Let, $\partial z/\partial t = v0$

Therefore, $dv/dt = v1\ e(j\omega t - \gamma z)(j\omega - \gamma\ v0)$ ------------------- (2)

Sub eqn (2) in (1)

$v1\ e(jwt - \gamma z)(j\omega - \gamma\ v0) = -e/mE1\ e(j\omega t - \gamma z)$

so, $v1(j\omega - \gamma\ v0) = -e/mE1$

$v1 = -E1e/m\ (j\omega - \gamma\ v0)$ ------------------- (3)

eqn (3) defines the change in electron velocity with respect to electric field, E1.

By law of conservation of electric charge, the continuity equation can be written as,

$\nabla . J = - \partial\rho / \partial t$ ---------------------- (4)

$\nabla = \partial / \partial x + \partial / \partial y + \partial / \partial z$

Consider only the z component

$\nabla = \partial / \partial z$ and $\nabla . J = - \partial J / \partial z$

Substitute for J from eq (c), we get

$\nabla . J = - \partial / \partial z(-J0 + J1\ e(j\omega t-\gamma z))$

$\nabla . J = - \gamma J1\ e(j\omega t-\gamma z)$ ----------------------- (5)

From eq (b)

$- \partial\rho / \partial t = - \partial / \partial t(\rho 0 + \rho 1\ e(j\omega t-\gamma z))$

$- \partial\rho / \partial t = - \rho 1\ j\omega e(j\omega t-\gamma z)$ ---------------------- (6)

Substitute eq (5) and (6) in eq (4)

$\nabla . J = - \partial\rho / \partial t$

$- \gamma J1\ e(j\omega t-\gamma z) = - \rho 1\ j\omega\ e(j\omega t-\gamma z)$

Therefore $- \gamma J1 = - \rho 1\ j\omega$

ie, $\rho 1 = \gamma J1 / j\omega$ -------------------------------- (7)

eqn (7) defines the change in charge density ρ with respect to current density J.

Consider eqn (e)

$J1 = \rho 0 v1 + \rho 1 v0$

Sub. for v1 from eqn (3) and ρ1 from eqn (7)

$J1 = \rho 0(-E1e/m\ (j\omega - \gamma\ v0)) + (\gamma J1 / j\omega)\ v0$

Thus, $J1 = -\rho 0 eE1\omega/m(j\omega - \gamma v0)(\omega + \gamma v0j)$

Multiplying Numerator and Denominator by v0

$J1 = -\rho 0 eE1\omega\ v0/m(j\omega - \gamma v0)(\omega + \gamma v0j)\ v0$

We have,

$J0 = -\rho 0\ v0$ and $Be = \omega 0\ v0$

Therefore,

$J1 = J0eE1Be/m(j\omega - \gamma v0)(\omega + \gamma v0j)$

Multiplying numerator and denominator with j

$J1 = J0eE1Bej/jm(j\omega - \gamma v0)(\omega + \gamma v0j)$

$J1 = J0eE1Bej/ m(-\omega - \gamma v0j)(\omega + \gamma v0j)$

$= J0eE1Bej/m((j\omega)2 - 2\omega\, \gamma v0j + (\gamma v0)2)$

$= J0eE1Bej/m(j\omega - \gamma v0)2$

$= J0eE1Bej/ v02m(j\omega/ v0 - \gamma)\ 2$

$= J0eE1Bej/ v02m(jBe - \gamma)\ 2$ [Since $\omega/ v0 = Be$]

The uniform velocity v0 is given by the eqn,

$v0 = \sqrt{}(2e\ v0/m)$

$v02 = 2e\ v0/m$

$e/ m\ v02 = 1/2\ v0$

Therefore **J1 = J0E1Bej/ 2v0(jBe - γ) 2**

When the electric field is uniform, J1 = i

Hence, **i = J0E1Bej/ 2v0(jBe - γ) 2**

This is the equation for convection current. This equation is also known as ‗electronic equation' as it determines the current induced by the electric field.

3.12 Axial Electric Field

Axial Electric Field in TWT

It defines how the electric field changes with convection current.

Let ‗L' be the inductance, ‗C' be the capacitance, ‗I' be the alternating current, ‗V' be the alternating voltage and ‗i' be the convection current. The current flowing is defined by the Equation,

$\partial \mathbf{I}/ \partial z = -C\partial V/ \partial t - \partial \mathbf{i}/ \partial z$ ------------------------- (1)

Let,

$\partial/ \partial z = -\gamma$ and $\partial/ \partial t = j\omega$

Therefore equation (1) becomes,

$-\gamma\ I = - j\omega CV + \gamma i$ --------------------------- (2)

The voltage across the electric field is defined as,

$\partial V/ \partial z = -L\partial \mathbf{I}/ \partial t$

$-\gamma V = - j\omega LI$

$I = \gamma V/ j\omega L$ -------------------------- (3)

Sub eq (3) in (1)

$-\gamma 2V/ j\omega L = -j\omega CV + \gamma i$

$\gamma 2V = -j\omega L(-j\omega CV + \gamma i)$

$= -CV\omega 2L - j\omega L\gamma i$

When the convection current, i = 0

$\gamma 2V = -CV\omega 2L$

$\gamma 2 = -C\omega 2L$

$\gamma = \sqrt{(-C\omega 2L)}$

$\gamma = j\omega\sqrt{(LC)}$

This equation represents propagation constant γ for an ideal transmission line.

ie, $\gamma 0 = j\omega\sqrt{(LC)}$

$\gamma 02 = -\omega 2LC$ --------------------------- (4)

When the convection current is not zero,

$\gamma 2V = -\omega 2LCV - \gamma ij\omega L$

$\gamma 2V + \omega 2LCV = -\gamma ij\omega L$

$V(\gamma 2 + \omega 2LC) = -\gamma ij\omega L$

$V = -\gamma ij\omega L/(\gamma 2 + \omega 2LC)$

Sub in eqn (4)

$V = -\gamma\omega Lij/(\gamma 2 - \gamma 02)$ ------------------------ (5)

Characteristic impedance,

$Zo = \sqrt{(L/C)}$

Multiplying both sides by γ0

$\gamma 0Zo = \gamma 0\sqrt{(L/C)}$

Sub for γ0 from eq (4)

$\gamma 0Zo = j\omega\sqrt{(LC)}.\ \sqrt{(L/C)}$

$= j\omega L$

Sub the value of γ0 Zo in eq (5)

$V = -i\gamma\ \gamma 0\ Z0\ /(\gamma 2 - \gamma 02)$

The electric field is defined as,

$Ez = -\nabla.\ V$

$Ez = -\partial V/ \partial z$ (we have, $\partial/ \partial z = -\gamma$)

$Ez = \gamma V$

Sub. for V.

Therefore, $Ez = -\gamma 2\ \gamma 0\ Z0\ \mathbf{i}/(\gamma 2 - \gamma 02)$

This equation is called circuit equation as it determines the electric field from the convection current, i.

3.13 Wave Modes in TWT

The solution for propagation constant ‗γ' in the electronic and circuit equations provides four distinct solutions. These four solutions represent four modes of propagation in a travelling wave tube. The values of four propagation constants are given by,

$\gamma 1 = -\beta e\ c\ \sqrt{(3/2)} + j\beta e\ [1 + (c/2)]$

$\gamma 2 = \beta e\ c\ \sqrt{(3/2)} + j\beta e\ [1 + (c/2)]$

$\gamma 3 = j\beta e\ [1 - c]$

$\gamma 4 = -j\beta e\ [1 - c3/4]$

The wave corresponding to γ1 is a forward wave and its amplitude grows exponentially with distance. The wave corresponding to γ2 is a forward wave whose amplitude decays exponentially. The wave corresponding to γ3 is a forward wave whose amplitude remains constant. The wave corresponding to γ4 is a backward wave whose amplitude remains constant. The growing wave propagates at a velocity slightly lower than electron velocity and energy flows from electron beam to travelling wave. The decaying wave propagates in the same manner as that of the growing wave but the energy flows from travelling wave to electron beam. The constant amplitude wave travels with a velocity higher than electron velocity and no energy transfer occurs. The backward wave propagates in negative z direction with a velocity higher than electron velocity.

3.14 Gain Consideration

The gain of the travelling wave tube is defined as,

$Ap = 10 \log |\text{output voltage/input voltage}|2$

$Ap = -9.54 + 47.3\ NC\ dB$

Where, N is length and C is gain parameter.

3.15 Magnetron Oscillators

A Magnetron Oscillator is used to generate high microwave power required in radar and communication systems. Magnetrons are cross field tubes in which the magnetic field and electric field are perpendicular to each other. All magnetrons consist of some form of anode and cathode operated in DC magnetic field which is perpendicular to DC electric field. Due to this cross field, electrons emitted from cathode moved in curved paths. If the magnetic field is strong enough the electrons will not arrive at the anode, but return to cathode itself. Thus the anode current is cut off. Magnetrons can be classified into three. They are

- **Split anode magnetron:**
  This type of magnetron uses negative resistance.
- **Cyclotron frequency magnetron:**
  This magnetron operates under the influence of synchronization between AC components of electric field and periodic oscillations of electrons in the direction parallel to the electric field.
- **Travelling wave magnetron:**
  This type of magnetron depends on the interactions of electrons with travelling electromagnetic field.

Cylindrical Magnetron

3.16 Cylindrical Magnetron Derivation

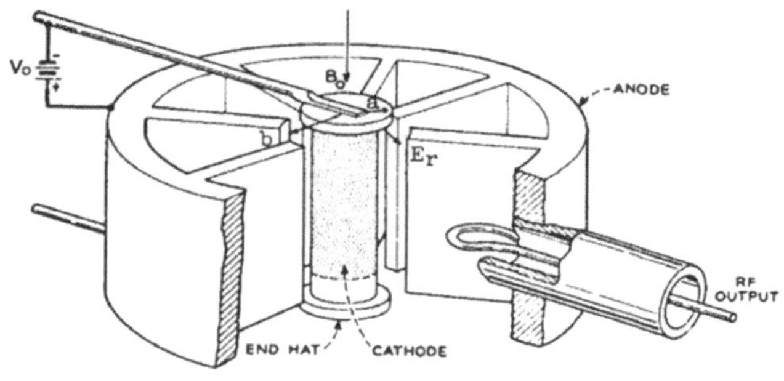

Schematic Diagram of Cylindrical Magnetron

Bo - Flux Density

a - radius of cathode cylinder

b - radius of anode cylinder

Equation of Electron Motion or Hull cut off Voltage equation

A charged particle in motion in a magnetic field of flux density B, experiences a force that is proportional to the charge, velocity, flux density and sine of angle between velocity and magnetic flux. The general equation for the motion of electron in terms of cylindrical co-ordinates is given as,

$r2(d\Phi/dt) = \frac{1}{2}\omega cr2 + \text{constant}$ ------------------ (1)

Where ωc is the cyclotron angular frequency given by,

$\omega c = eBz/m$

e – charge of electron

Bz – magnetic flux density

M – mass of electron

Let, $r = a$ and $d\Phi/dt = 0$

Therefore, Equation (1) becomes

$\frac{1}{2}\omega ca2 + \text{Constant} = 0$

$\text{Constant} = -\frac{1}{2}\omega ca2$ -------------------- (2)

Substitute equation (2) in (1)

$r2(d\Phi/dt) = \frac{1}{2}\omega cr2 - \frac{1}{2}\omega ca2 = \frac{1}{2}\omega c(r2 - a2)$

$d\Phi/dt = \omega c(r2 - a2)/2r2 = \frac{1}{2}\omega c\ (1 - (a2/\ r2))$ --------------- (3)

The KE of electrons is given by,

$\frac{1}{2}mv2 = eV$

$v2 = 2eV/m$

Considering r and Φ components of electron velocity, the above equation can be represented as,

$vr2 + v\Phi 2 = 2eV/m$

$(dr/dt)2 + (r\ d\Phi/dt)2 = 2eV/m$ --------------- (4)

Let r = b

V = Vo

dr/dt = 0

Therefore the above equation (4) becomes,

(bdΦ/dt)2 = 2eVo /m ------------------- (5)

Substituting dΦ/dt from eq (3)

b2(ωc/2 (1- (a2/ r2)))2 = 2eVo /m

Substituting r = b

b2(ωc (1- (a2/ b2)))2 = 2eVo /m

b2ωc/4(1- (a2/ b2))2= 2eVo /m

b2ωc2(1- (a2/ b2))2= 8eVo /m

Substituting ωc = eBz/m

b2(eBz/m)2(1- (a2/ b2))2= 8eVo /m

b2e2/m2 (1- (a2/ b2))2 Bz2 = 8eVo /m

$$B_z = \sqrt{\frac{8V_o m}{eb^2}\frac{1}{(1-(\frac{a}{b})^2)^2}}$$

This magnetic field intensity is known as cut off flux density, Boc

$$B_{oc} = \sqrt{\frac{8V_o m}{eb^2}\frac{1}{(1-(\frac{a}{b})^2)^2}}$$

Voc = b2e Bz2 (1 –(a/b)2)2/8m

This equation is known as ‗Hull cut-off voltage equation'

Cyclotron Angular Frequency:

The cyclotron angular frequency of cylindrical magnetron is given by,

ωc = eB/m

The period of oscillation is given by,

T = 2 π/ ωc

T = 2 πm/eB

The phase shift is given by,

$\Phi = 2\pi m/N$

Where, N is the number of cycles.

Output Power and Efficiency

The efficiency and output power of the magnetron depend on the structure and power supply. The equivalent circuit of magnetron is,

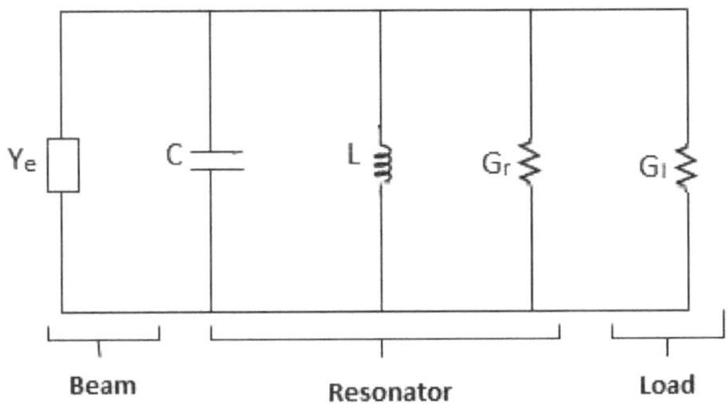

Ye – admittance

C – Capacitance

L – Inductance

Gr – Resonator Conductance

Gl – Load Conductance

The unloaded quality factor is given by,

$Q_{un} = \omega_o C/G_r$

The loaded quality factor is given by,

$Q_l = \omega_o C/G_r + G_l$

The external quality factor,

$Q_{ext} = \omega_o C/G_l$

The circuit efficiency, $\eta_c = 1/(1 + (Q_{ext}/Q_{un}))$

Electronic efficiency, $\eta_e = (V_oI_o - P_{lost}) / V_oI_o$

Where VO – Anode Voltage

Io – Anode Current or Beam Current

Plost – Power lost in the anode.

3.17 Coaxial Magnetron

The coaxial magnetron is composed of an anode resonator structure surrounded by an inner-single, high-Q cavity operating in the TE_{011} mode as shown in Figure :

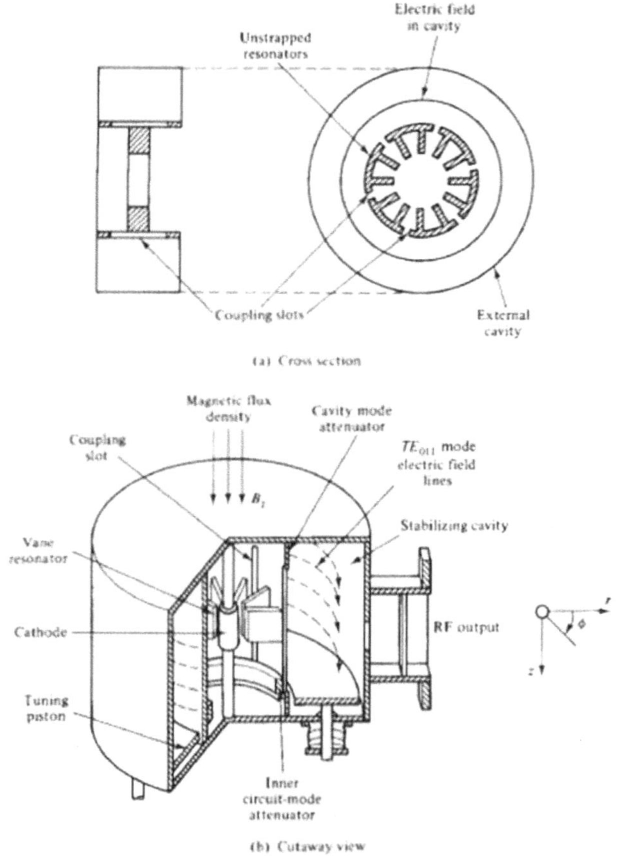

Figure: Schematic diagram of a coaxial magnetron

The slots in the back walls of alternate cavities of the anode resonator structure tightly couple the electric fields in these resonators to the surrounding cavity. In the π-mode operation, the electric fields in every other cavity are in phase, and so they couple in the same direction into the surrounding cavity. As a result, the surrounding coaxial cavity stabilizes the magnetron in the desired rr-mode oepration. In the desired TE_{011} mode, the electric fields follow a circular path within the cavity and reduce to zero at the walls of the cavity. Current flow in the TE_{011} mode is in the walls of the cavity in circular paths about the axis of the tube. The undesired modes are damped out by the attenuator within the inner slotted cylinder near the ends of the coupling slots. The tuning mechanism is simple and reliable. As the straps are not required, the anode resonator for the coaxial magnetron can be larger and less complex than for the conventional strapped magnetron. Thus cathode loading is lower, and voltage gradients are reduced.

In the desired TEoI I mode, the electric fields follow a circular path within the cavity and reduce to zero at the walls of the cavity. Current flow in the TE_{011} mode is in the walls of the cavity in circular paths about the axis of the tube. The undesired modes are damped out by the attenuator within the inner slotted cylinder near the ends of the coupling slots. The tuning mechanism is simple and reliable. As the straps are not required, the anode resonator for the coaxial magnetron can be larger and less complex than for the conventional strapped magnetron. Thus cathode loading is lower, and voltage gradients are reduced.

UNIT - IV

SIGNAL DEGRADATION IN OPTICAL FIBERS

Attenuation - Absorption losses - Scattering losses - Bending Losses - Core and Cladding losses - Signal Distortion in Fibers - Intermodal delay - Intramodal dispersion - Factors contributing to dispersion - Group Delay - Material Dispersion - Wave guide Dispersion - Basics of semiconductor physics - LED - Structures - Light source materials - Quantum efficiency and LED power - LASER diodes.

4.1 Attenuation

What is Attenuation : Different Types & Its Causes

The attenuation is a telecommunication word which refers to reduction within signal strength. This can occur while transmitting signals over lengthy distances. It can be calculated in dB (decibels) in terms of voltage. The function of this is quite opposite to amplification when a signal is transmitted from one place to another place. Once the signal attenuation is extremely high, and then it turns into incoherent. So, most of the networks use repeaters for increasing the signal strength at normal intervals.

What is Attenuation?

Attenuation meaning is the reduction of signal strength and it can occur in any kind of signal like analog otherwise digital. In some cases, it can be called attenuation loss; because this is a normal effect of a signal while transmitting over long distances. In some cables like conventional or FOCs (fiber optic cables), this can be identified in terms DBs (decibels) for each foot, kilometer, or thousand foot, etc. The cable efficiency is high when the attenuation for each unit distance is less.

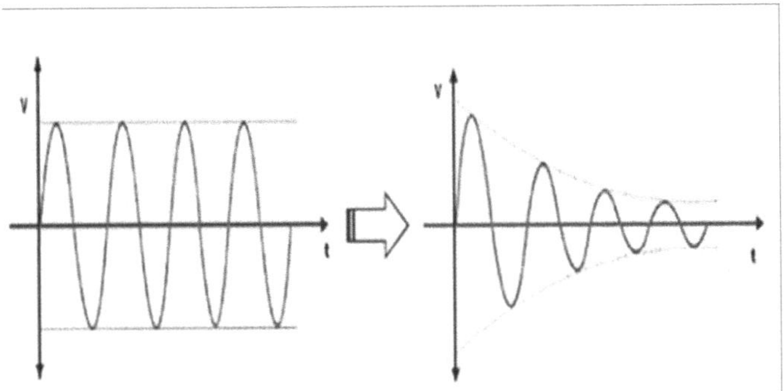

Attenuation in signals

When it is required to send out signals over long distances through any cable, then one (or) more repeaters must be included with the length of the cable. Because repeaters play a key role in enhancing the strength of the signal to conquer this. So this enhances the highest range of achievable communication.

Causes of Attenuation

This can be occurred in wired as well as wireless transmissions due to signaling issues. There are several examples are there in digital network circuitry & telecommunications. This can be occurred due to the following reasons.

Transmission Medium

An electromagnetic field can occur around the transmission once all the signals broadcasted down, then energy losses will occur in the downside of the cable based on the length & frequency of the cable.

Crosstalk

Crosstalk from nearby cable can cause this within cables like conductive metal or copper.

Connectors & Conductors

Attenuation can take place when a signal flows across dissimilar conductive standards & connector surfaces. The circuits can be attenuated by using repeaters for signal boosting via amplification. When copper conductors are used, then the high-frequency signal and additional attenuation can occur with a length of cable. Present communications use HFs (high-frequencies), thus the mediums which have a smooth-attenuation in all frequencies like fiber optics are employed instead of normal copper circuits.

Noise

Additional noise on N/Ws (networks) such as RFs (radio frequencies), leakage in wires, electrical currents can interfere by the signal to cause this. If the noise is more, then this will be more.

Physical Environs

Physical environs which include installation of improper wiring, wall barriers, the temperature can alter the transmission, then attenuation can be caused.

Travel Distance

When the transmission in a cable travels for long distances like from source (current place) to destination (connection supplier), then it experiences more noise while traveling.

Different Types

There are different types of attenuations which include deliberate, automatic, and environmental.

- **Deliberate**
  This kind of attenuation can happen wherever a volume control can be used to reduce the level of sound over consumer electronics.
- **Automatic**
  This kind of attenuation is used to stop the distortion of sound in audio equipment and TVs by detecting automatic level to activate attenuation circuits.
- **Environmental**
  This kind of attenuation relates to loss of signal power because of the transmission medium, whether that can be connected to copper wire, fiber optic or wireless.

Attenuation in Optical Fiber

Attenuation can occur to any kind of signal like fiber, copper, satellite, fiber, etc. In the Fiber signal, it travels on HF (high-frequency) wavelength light which can be protected by glass tubes. When light is opposed to noise sources such as RFs, electricity, the attenuation rate of fiber connections has extremely low.

The proper functioning of optical data link mainly depends on the light which is changed t reach the receiver by sufficient power to be de-modulated properly. This is the drop within the light signal power while it is transmitted. This can be occurred due to some passive media components which include connectors, splices of cables, and cables.

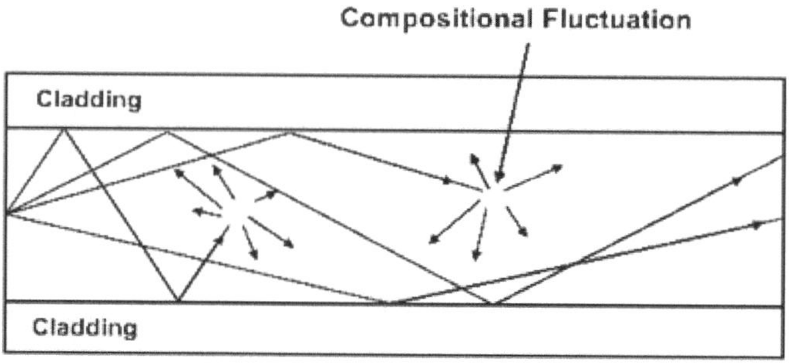

Attenuation in optical fiber

Even though this is considerably lower for this cable when compared with other media. In fiber optic, the transmission can be done in two modes like single-mode and multi-mode. But, attenuation can occur in both the transmission modes. So this can be avoided by maintaining enough light in an optical data link.

The size of single-mode fiber is very small and the internal light reflection can travel through only a single layer. The interfacing of this optic mainly uses laser lights and generates light in a single wavelength. The bandwidth of this fiber is high & carries signals for long distances.

The size of multimode fiber is large and the internal light reflection can travel through a multi-wavelength. The interfacing of this optic mainly uses LEDs & generates light in different wavelengths and causes signal dispersion.

When light reflection travels within the fiber core then it emits into the cladding, higher-order mode loss results. Mutually these issues will stop the distance of transmission in multimode compared with single-mode. As the maximum transmission distance increases, it can result in a signal loss & causes variable transmission.

Attenuation Coefficient

The attenuation coefficient of FOC (fiber optic cable) is one of the most significant parameters. In a huge amount, the distance of relay can be decided within the optical transmission.

The attenuation coefficient of fiber can be 0.36dB/km in wavelength of 1310nm as well as 0.22dB/km in wavelength of 1550nm.

Attenuation Measurement

Generally, the amount of attenuation can be expressed in dB (decibels) units.

If the signal power ‗Ps' at the source of a circuit & the signal power ‗Pd' is at the destination, followed by Ps is greater than Pd. The power attenuation ‗Ap' in dB can be denoted by using the following attenuation formula

Ap = 10 log10 * (Ps/Pd)

Attenuation in terms of voltage can also be expressed. If the voltage attenuation is ‗Av' in dB, source signal voltage is ‗Vs' & the destination signal voltage ‗Vd' then the equation will be

Av = 20 log10 * (Vs/Vd)

Thus, this is all about an overview of attenuation in fiber optic cable. It is a reduction of signal potency and can be calculated in dB. It reduces the maximum speed connections accessible because of the requirement of numerous recur transmissions.

4.2 Absorption Losses

Absorption loss is related to the material composition and fabrication process of fiber. Absorption loss results in dissipation of some optical power as hear in the fiber cable. Although glass fibers are extremely pure, some impurities still remain as residue after purification. The amount of absorption by these impurities depends on their concentration and light wavelength.

a. Intrinsic absorption

Intrinsic absorption in the ultraviolet region is caused by electronic absorption bands. Basically, absorption occurs when a light particle (photon) interacts with an electron and excites it to a higher energy level. The main cause of intrinsic absorption in the infrared region is the characteristic vibration frequency of atomic bonds. In silica glass, absorption is caused by the vibration of silicon-oxygen (Si-O) bonds. The interaction between the vibrating bond and the electromagnetic field of the optical signal causes intrinsic absorption. Light energy is transferred from the electromagnetic field to the bond.

b. Extrinsic absorption

Extrinsic absorption is much more significant than intrinsic Caused by impurities introduced into the fiber material during manufacture – Iron, nickel, and chromium Caused by transition of metal ions to higher energy level Modern fabrication techniques can reduce impurity levels below 1 part in 1010. For some of the more common metallic impurities in silica fibre the table shows the peak attenuation wavelength and the attenuation caused by an impurity concentration of 1 in 109

OR

Radiative losses

Radiative losses also called bending losses, occur when the fibre is curved. There are two types of radiative losses: Micro bending losses. Macro bending losses.

OR

Scattering loss

Basically, scattering losses are caused by the interaction of light with density fluctuations within a fiber. Density changes are produced when optical fibers are manufactured.

Linear Scattering Losses: Linear scattering occurs when optical energy is transferred from the dominant mode of operation to adjacent modes. It is proportional to the input optical power injected into the dominant mode. o Linear scattering is divided into two categories: Mie scattering and Rayleigh scattering.

Non- Linear Scattering Losses: Scattering loss in a fiber also occurs due to fiber non-linearity's i.e. if the optical power at the output of the fiber does not change proportionately with the power change at the input of the fiber, the optical fiber is said to be operating in the non-linear mode. Non-Linear scattering is divided into two categories: Stimulated Raman Scattering and Stimulated Brillouin Scattering.

OR

Dispersion loss

Dispersion is a measure of the temporal spreading that occurs when a light pulse propagates through an optical fiber. Dispersion is sometimes referred to as delay distortion in the sense that the propagation time delay causes the pulse to broaden.

4.3 Scattering Losses

Basically, scattering losses are caused by the interaction of light with density fluctuations within a fiber. Density changes are produced when optical fibers are manufactured. During manufacturing, regions of higher and lower molecular density areas, relative to the average density of the fiber, are created.

Light traveling through the fiber interacts with the density areas as shown in Light is then partially scattered in all direction.

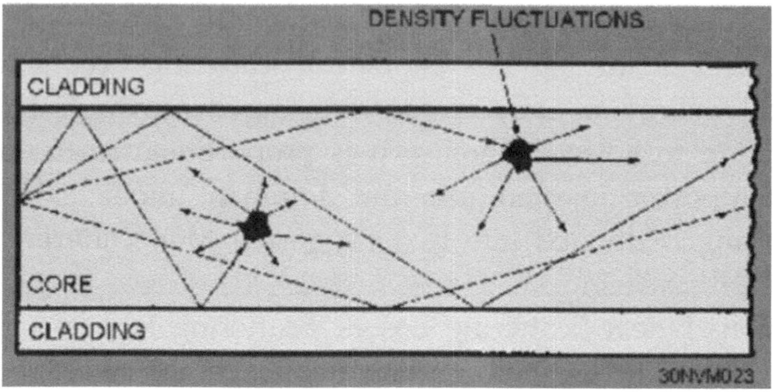

Scattering Process

In commercial fibers operating between 700-nm and 1600-nm wavelength, the main source of loss is called **Rayleigh scattering.** As the wavelength increases, the loss caused by Rayleigh scattering decreases. If the size of the defect is greater than one-tenth of the wavelength of light, the scattering mechanism is called **Mie scattering.**

Linear Scattering Losses

Rayleigh Scattering

It occurs because the molecules of silicon dioxide have some freedom when adjacent to one another. Thus, setup at irregular positions and distances with respect to one another when the glass is rapidly cooled during the final stage of the fabrication process. Those structural variations are seen by light as variations in the refractive index, thus causing the light to reflect – that is, to scatter – in different directions

Rayleigh scattering is a scattering of light by particles much smaller than the wavelength of the light, which may be individual atoms or molecules. Rayleigh scattering is a process in which light is scattered by a small spherical volume of variant refractive index, such as a particle, bubble, droplet, or even a density fluctuation.

As light travels in the core, it interacts with the silica molecules in the core. Rayleigh scattering is the result of these elastic collisions between the light wave and the silica molecules in the fiber. Rayleigh scattering accounts for about 96 percent of attenuation in optical fiber

Causes of Rayleigh Scattering

- It results from non-ideal physical properties of the manufactured fiber.
- It results from inhomogeneities in the core and cladding. Because of these inhomogeneities problems occur like –
  1. Fluctuation in refractive index
  2. Density and compositional variations.

Minimizing of Rayleigh Scattering

Rayleigh scattering is caused due to compositional variations which can be reduced by improved fabrication.

Equation of Rayleigh Scattering

Light scattering can be divided into three domains based on a dimensionless size parameter, α which is defined as

$$\mathbf{A = \pi Dp / \lambda}$$

where, πDp is the circumference(The boundary line of a circle) of a particle and λ is the wavelength of incident radiation. Based on the value of α, these domains are:

$\alpha<<1$: Rayleigh scattering (small particle compared to wavelength of light) $\alpha\approx1$: Mie scattering (particle about the same size as wavelength of light)

Mie Scattering

Non perfect cylindrical structure of the fiber and imperfections like irregularities in the core-cladding interface, diameter fluctuations, strains and bubbles may create linear scattering which is termed as Mie scattering

Mie scattering is a scattering of light by particles approximately equal to the wavelength of the light, which may be individual atoms or molecules.

Causes of Mie Scattering

- Occurred due to inhomogeneities in the composition of silica. (i.e. inhomogeneities in the density of SiO2)
- Irregularities in the core-cladding interface, Difference in core cladding refractive index, Diameter fluctuations
- Due to presence of strains and bubbles.

The scattering caused by such inhomogeneities is mainly in the forward direction depending upon the fiber material, design and manufacture.

Minimizing of Mie Scattering

Mie scattering is mainly caused by inhomogeneities which can be minimized by

- Removing imperfection due to glass manufacturing process
- Carefully controlled extrusion(To push or thrust out) and coating of the fiber

Both Mie and Rayleigh scattering are considered elastic scattering (elastic scattering is also called Linear scattering) processes, in which the energy (and thus wavelength and frequency) of the light is not substantially changed.

Nonlinear Scattering Losses

Specially at high optical power levels scattering causes disproportionate attenuation, due to non linearbehavior. Because of this non linear scattering the optical power from one mode is transferred in either the forward or backward direction to the same, or other modes, at different frequencies. The two dominant types of non linear scattering are :

- Stimulated Brillouin Scattering and
- Stimulated Raman Scattering.

Stimulated Brillouin Scattering

This is defined as the modulation of light through thermal molecular vibration within the fiber. The scattered light contains upper and lower side bands along with incident light frequency. An incident photon produces a scattered photon as well as a photon of acoustic frequency. The frequency shift is maximum in the backward direction and it is reduced to zero in the forward direction. The threshold optical power for Brillion scattering is proportional to

$\mathbf{d^2\lambda^2\alpha B}$

Stimulated Raman Scattering

Here, the scattered light consists of a scattered photon and a high frequency optical photon. Further, this occurs both in the forward and backward direction in the optical fiber. The threshold optical power for Raman scattering is about three orders of magnitude higher than the Brillouin threshold for the given fiber, The threshold optical power for Raman scattering is proportional to

$\mathbf{d^2\lambda^2\alpha R}$

4.4 Bending Losses

Optical fibers suffer radiation losses at bends or curves on their paths. This is due to the energy in the evanescent field at the bend exceeding the velocity of light in the cladding and hence the guidance mechanism is inhibited, which causes light energy to be radiated from the fiber. An illustration of this situation is shown in Figure 2.5. The part of the mode which is on the outside of the bend is required to travel faster than that on the inside so that a wavefront perpendicular to the direction of propagation is maintained.

Hence, part of the mode in the cladding needs to travel faster than the velocity of light in that medium. As this is not possible, the energy associated with this part of the mode is lost through radiation. The loss can generally be represented by a radiation attenuation coefficient which has the form:

$$\alpha_r = c_1 \exp(-c_2 R)$$

Where, R is the radius of curvature of the fiber bend and c1, c2 are constants which are independent of R. Furthermore, large bending losses tend to occur in multimode fibers at a critical radius of curvature Rc which may be estimated from:

$$R_c \simeq \frac{3 n_1^2 \lambda}{4\pi (n_1^2 - n_2^2)^{\frac{3}{2}}}$$

It may be observed from the expression given in above equation that potential macrobending losses may be reduced by:

- Designing fibers with large relative refractive index differences;
- Operating at the shortest wavelength possible.

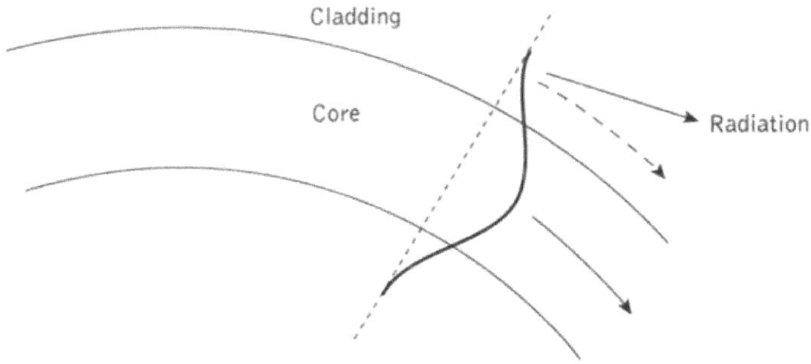

An illustration of the radiation loss at a fiber bend

The above criteria for the reduction of bend losses also apply to single-mode fibers. One theory, based on the concept of a single quasi-guided mode, provides an expression from which the critical radius of curvature for a single-mode fiber Rcs can be estimated as

$$R_{cs} \simeq \frac{20\lambda}{(n_1 - n_2)^{\frac{3}{2}}}\left(2.748 - 0.996\frac{\lambda}{\lambda_c}\right)^{-3}$$

Where, λc is the cutoff wavelength for the single-mode fiber. Hence again, for a specific single-mode fiber (i.e. a fixed relative index difference and cutoff wavelength), the critical wavelength of the radiated light becomes progressively shorter as the bend radius is decreased.

4.5 Core and Cladding Losses

Since the core and cladding have different indices of refraction hence they have different attenuation coefficients α1 and α2 respectively.

$$\alpha(r) = \alpha_1 + (\alpha_2 - \alpha_1)\frac{n^2(0) - n^2(r)}{n^2(0) - n_2^2}$$

For step index fiber, the loss for a mode order (v, m) is given by,

$$\alpha_{vm} = \alpha_1 \frac{P_{core}}{P} + \alpha_2 \frac{P_{cladding}}{P}$$

For low-order modes, the expression reduced to

$$\alpha_{vm} = \alpha_1 + (\alpha_2 + \alpha_1) \frac{P_{cladding}}{P}$$

where, $\frac{P_{core}}{P}$ and $\frac{P_{cladding}}{P}$ are fractional powers.

For graded index fiber, loss at radial distance is expressed as, The loss for a given mode is expressed by,

$$\alpha_{Graded\ Index} = \frac{\int_0^\infty \alpha(r)\, P(r)\, r\, dr}{\int_0^\infty P(r)\, r\, dr}$$

where, P(r) is power density of that model at radial distance r.

Signal Distortion in Optical Waveguide

The pulse gets distorted as it travels along the fiber lengths. Pulse spreading in fiber is referred as dispersion. Dispersion is caused by difference in the propagation times of light rays that takes different paths during the propagation. The light pulses travelling down the fiber encounter dispersion effect because of this the pulse spreads out in time domain. Dispersion limits the information bandwidth. The distortion effects can be analyzed by studying the group velocities in guided modes.

4.6 Signal Distortion in Fibers

What is meant by signal distortion?

It means the alteration of the waveform of an information-bearing signal, such as an audio signal representing sound or a video signal representing images, in an electronic device or communication channel.

Why signal gets distorted while propagating on optical fiber?

The loss of optical fiber is wavelength dependent. That makes a function of frequency.This is called the attenuation of the fiber.

Different wavelengths and different modes of the optical fiber travel with different speedsand therefore have delay which is wavelength dependent. This phenomenon is called thedispersion.

So, in general the signals get distorted on an optical fiber because differentfrequency components undergo different attenuation and different delays.For an optical communication link the highest data rate is about 10GHz. This translatesapproximately to a wavelength range of 0.1nm. For all practical purposes, the loss of theoptical fiber can be assumed constant over a bandwidth of 0.1nm. Therefore, we note thatuniform amplitude response condition is well satisfied for the optical fiber. The signaldepending upon the loss may reduce in amplitude but there is no distortion of the signaldue the loss.

4.7 Intermodal Delay

Intermodal dispersion (also called modal dispersion) is the phenomenon that the group velocity of light propagating in a multimode fiber (or other waveguide) depends not only on the optical frequency (→ chromatic dispersion) but also on the propagation mode involved.

Figure shows a numerical simulation, where a 200-fs ultrashort pulse is launched into a 50 cm long multimode fiber such that multiple modes are excited. After the fiber, the corresponding contributions appear at different times due to different group velocities of the modes. The fundamental mode comes first, as it is the fastest.

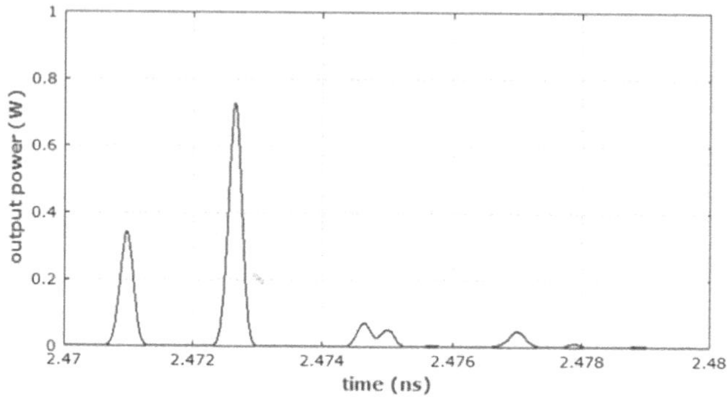

Output power versus time for a 200-fs input pulse injected into a 50 cm long multimode fiber. The numerical simulation has been done with the RP Fiber Power software.

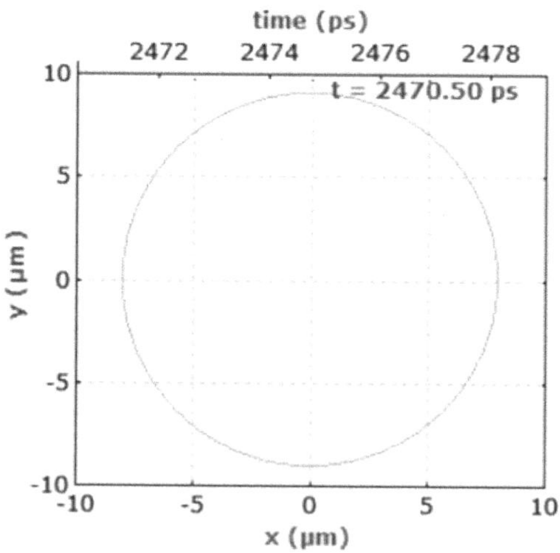

Time-dependent output beam profile for the same situation as in Figure 1

The strength of intermodal dispersion can be quantified as the differential mode delay (DMD). It depends strongly on the refractive index profile of the fiber in and around the fiber core. For example, for a step-index profile the higher-order modes have lower group velocities, and this can lead to differential group delays of the order of 10 ps/m = 10 ns/km. It is then hardly possible to realize data rates of multiple Gbit/s in an fiber-optic link with a kilometer length.

In systems for optical fiber communications based on multimode fibers, intermodal dispersion can severely limit the achievable data transmission rate (bit rate). In order to avoid strong signal distortion, it is usually necessary to keep the pulses long enough to maintain a reasonable temporal overlap of components from different modes, and this unavoidably sets a limit on the data rate.

The natural way of eliminating intermodal dispersion is to use fiber links based on single-mode fibers: if there is only one propagation mode available (disregarding possible polarization mode dispersion and cladding modes), there cannot be difference between propagation times. However, intermodal dispersion can also be minimized by using multimode fibers with a parabolic refractive index profile, where intermodal dispersion is minimized.

4.8 Intramodal Dispersion

Dispersion

Dispersion is the broadening of actual time-width of the pulse due to material properties and imperfections.

As pulse travels down the fiber, dispersion causes pulse spreading. This limits the distance travelled by the pulse and the bit rate of data on optical fiber.

In a fiber three distinct types of distortion are observed:

Intramodal Dispersion

Pulse broadening within a single mode is called as intramodal dispersion or chromatic dispersion.

Since this phenomenon is wavelength dependent and group velocity is a function of wavelength, it is also called as group velocity dispersion (GVD).

The two main causes of intramodal dispersion are as follows:

a. **Material dispersion:**
   - It is the pulse spreading due to the dispersive properties of material.
   - It arises from variation of refractive index of the core material as a function of wavelength.
   - Material dispersion is a property of glass as a material and will always exist irrespective of the structure of the fiber.

b. **Waveguide dispersion:**
   - It occurs because a single mode fiber confines only about 80% of the optical power to the core.
   - Dispersion thus arises since the 20% light propagating in the cladding travels faster than light confined to the core.
   - The amount of waveguide dispersion depends on the structure of the fiber and can be varied by altering the parameters such as NA, core radius etc.

4.9 Factors Contributing to Dispersion

1. Laser spectral width, modulation method, and frequency chirp. Lasers with wider spectral widths and chirp have shorter dispersion limits. It is important to refer to manufacturer specifications to determine the total amount of dispersion that can be tolerated by the lightwave equipment.

2. The wavelength of the optical signal. Chromatic dispersion varies with wavelength in a fiber. In a standard non-dispersion shifted fiber (NDSF G.652), chromatic dispersion is near or at zero at 1310 nm. It increases positively with increasing wavelength and increases negatively for wavelengths less than 1310 nm.
3. The optical bit rate of the transmission laser. The higher the fiber bit rate, the greater the signal distortion effect.
4. The chromatic dispersion characteristics of fiber used in the link. Different types of fiber have different dispersion characteristics.
5. The total fiber link length, since the effect is cumulative along the length of the fiber.
6. Any other devices in the link that can change the link's total chromatic dispersion including chromatic dispersion compensation modules.
7. Temperature changes of the fiber or fiber cable can cause small changes to chromatic dispersion. Refer to the manufacturer's fiber cable specifications for values.

4.10 Group Delay

The group delay (Tg) of an optical element (e.g. a dielectric mirror or a piece of optical fiber) is essentially the time delay experienced by a short (but not too short and not too intense) light pulse for propagation through that element. That time delay generally depends on the optical wavelength. More precisely, the group delay is defined as the derivative of the change in spectral phase with respect to the angular frequency:

$$T_g = \frac{\partial \varphi}{\partial \omega}$$

The group delay has the units of a time (e.g. picoseconds) and generally (in dispersive media) depends on the optical frequency (→ group delay dispersion, chromatic dispersion) and possibly on the polarization state (→ polarization mode dispersion) and the optical mode (→ intermodal dispersion) in case of a waveguide.

As a simple example, for propagation over a distance d in vacuum we have $\varphi = 2\pi d / \lambda = \omega d / c$, so that the resulting group delay is d / c.

For linear propagation of a narrow-band optical pulse with a simple temporal and spectral shape, the group delay is the time delay which the pulse maximum (i.e., the maximum of the temporal intensity profile) experiences when propagating through the optical element. For broadband optical pulses, and particularly in situations where nonlinearities affect the propagation, the situation can be more sophisticated, leading to a time delay which can deviate from the group delay. Inappropriate interpretations of the group delay can then cause confusion. There are even situations where the temporal pulse shape undergoes a complicated evolution, e.g. with multiple intensity peaks which can vary in strength such that the global pulse maximum can suddenly shift its temporal location.

The group velocity of light in a medium is the inverse of the group delay per unit length. For a piece of optical material with a certain length, the group delay is the length divided by the group velocity. Note that for common transparent optical materials (also for solid-state laser crystals and optical fibers), the group velocity can significantly deviate from the phase velocity. For example, one meter of fused silica bulk material causes a group delay of 4.879 ns at 1550 nm, whereas from the phase velocity one would (wrongly) calculate 4.817 ns. For a shorter wavelength like 400 nm, this discrepancy is larger: the group delay is 5.049 ns instead of 4.878 ns. In optical fibers, the group delay is further modified by dopants of the fiber core and by the effect of waveguide dispersion.

For an optical resonator, the group delay (and not the phase delay) for one round trip determines the resonator mode spacing, also called the free spectral range.

Measurement of Group Delay

The group delay of an optical element can be measured in various ways. A conceptually most direct method is based on measuring the arrival times of ultrashort pulses e.g. with fast photodetectors. Similarly, one can measure a phase change of intensity-modulated light, with its optical power detected before and after the optical element (phase shift method).

There are more sophisticated and far more powerful interferometric methods, e.g. based on white light interferometry, which allow the measurement of wavelength-resolved spectral phase changes and the group delay with a precision of a few femtoseconds.

Differential Group Delay

In some situations, a quantity of primary interest is a differential group delay, i.e., a difference of two different group delays. For example, in a birefringent optical medium there is generally a difference in group delay between two polarization directions. That can be a problem for nonlinear frequency conversion with ultrashort pulses in nonlinear crystals, where the differential group delay leads to a temporal walk-off of the interacting pulses. See also the article on polarization mode dispersion, which is also observed even in nominally not birefringent fibers. In some cases, a differential group delay is compensated by inserting a piece of birefringent material which provides the opposite differential group delay.

Another example is a differential group delay for two pulses or for two optical telecom signals with different central wavelengths propagating through an optical fiber. That kind of differential group delay arises from the group velocity dispersion of the fiber.

4.1 Material Dispersion

Material dispersion is noted in the centre of the of the optical fiber (the core) due to its build. Optical fibers constitute glass fibers which are in turn materials that are formless.

Amorphous materials have no pattern or form in an atomic level. Two rays of light that share the same wavelength and transfer through the same path length can be conveyed in different areas of the core.

Due to this formless structure, these different areas will not bear 100% the same atomic level make-up as well as RI. Differences in rate are driven by atomic level differences in structure and likewise differences occur from differences between the compositions in the GL core. This type of dispersion is seen in both single mode as well as multimode fiber types.

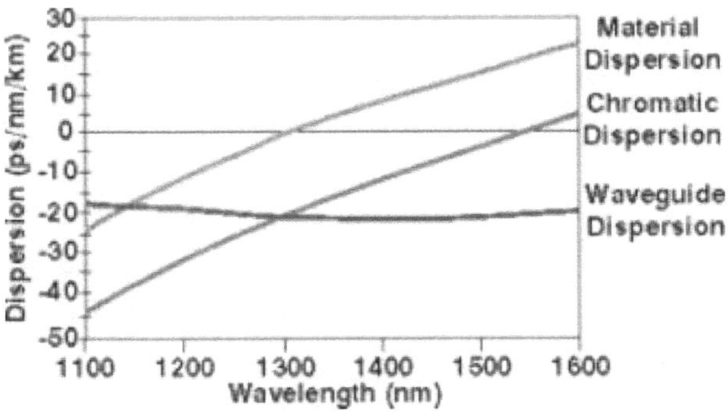

Material, Chromatic and Waveguide dispersions per different wavelengths

Dispersion of the material leads to one out of three basic borders on bandwidth and conveyance distance. It is not possible to zero or cover material dispersion as in the case of modal and chromatic dispersion types.

4.12 Wavc Guide Dispersion

The term —Waveguide dispersion‖ is the distribution of energy occurring from different speeds or RIs in the core and cladding regions of the fiber. It's not possible to expel waveguide dispersion as in the case of modal and chromatic dispersion types.

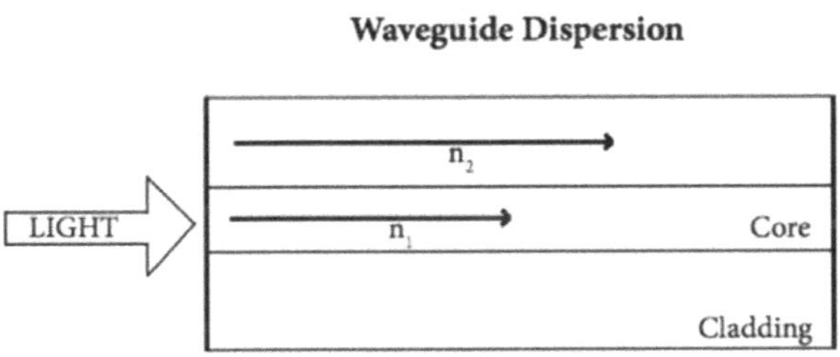

Waveguide dispersion

Polarization Mode Dispersion

When the light is defined as an energy wave or energy region, it possesses 2 mutually perpendicular axes, the E (also known and electromotive force) and H, (also known as the magneto motive force).

The moment the energy inside these two axes transfers at different speeds in a fiber this leads to Polarization Mode Dispersion. PMD's significance raises in levels that exceed 2.5 Gbps at long transmission distances.

Due to the fact that these rates and distances exceed much the level of data networks, PMD isn't an issue in these types of networks. Still, PMD can be vital for telephone, CATC and networks of longer distance.

It's not possible to zero or cover PMD as in the case of modal and chromatic dispersion types. Essentially, PMD is more significant in old fibers that were installed at least 15 years ago than newer optical fibers that were installed after 2000.

4.13 Basics of Semiconductor Physics

A semiconductor is a substance whose resistivity lies between the conductors and insulators. The property of resistivity is not the only one that decides a material as a semiconductor, but it has few properties as follows.

Semiconductors have the resistivity which is less than insulators and more than conductors.

Semiconductors have negative temperature co-efficient. The resistance in semiconductors, increases with the decrease in temperature and vice versa.

The Conducting properties of a Semi conductor changes, when a suitable metallic impurity is added to it, which is a very important property.

Semiconductor devices are extensively used in the field of electronics. The transistor has replaced the bulky vacuum tubes, from which the size and cost of the devices got decreased and this revolution has kept on increasing its pace leading to the new inventions like integrated electronics. The following illustration shows the classification of semiconductors.

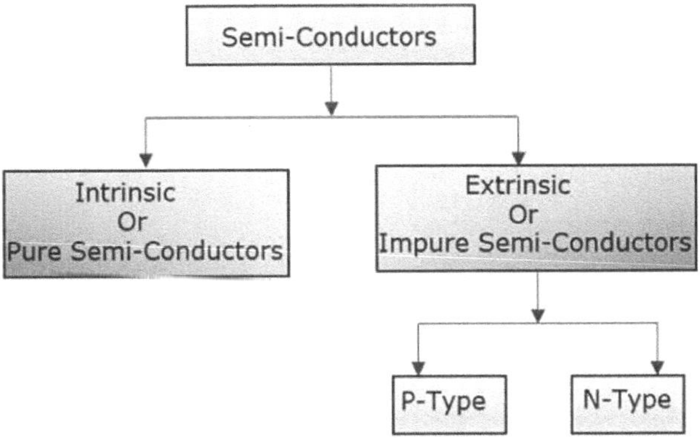

Conduction in Semiconductors

After having some knowledge on the electrons, we came to know that the outermost shell has the valence electrons which are loosely attached to the nucleus. Such an atom, having valence electrons when brought close to the other atom, the valence electrons of both these atoms combine to form ―Electron pairs‖. This bonding is not so very strong and hence it is a Covalent bond.

For example, a germanium atom has 32 electrons. 2 electrons in first orbit, 8 in second orbit, 18 in third orbit, while 4 in last orbit. These 4 electrons are valence electrons of germanium atom. These electrons tend to combine with valence electrons of adjoining atoms, to form the electron pairs, as shown in the following figure.

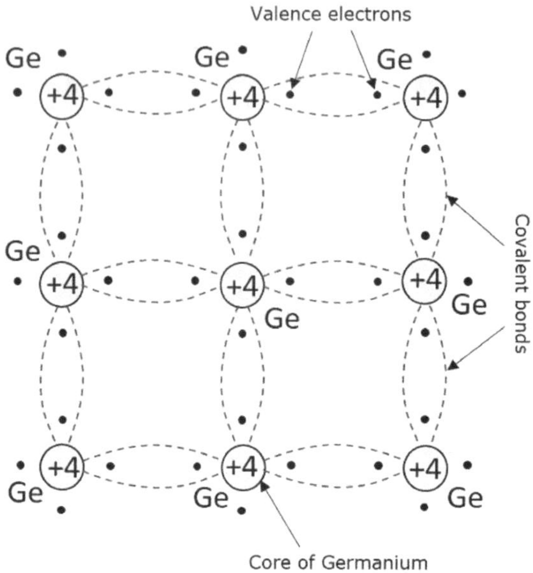

Creation of Hole

Due to the thermal energy supplied to the crystal, some electrons tend to move out of their place and break the covalent bonds. These broken covalent bonds, result in free electrons which wander randomly. But the moved away electrons creates an empty space or valence behind, which is called as a hole.

This hole which represents a missing electron can be considered as a unit positive charge while the electron is considered as a unit negative charge. The liberated electrons move randomly but when some external electric field is applied, these electrons move in opposite direction to the applied field. But the holes created due to absence of electrons, move in the direction of applied field.

Hole Current

It is already understood that when a covalent bond is broken, a hole is created. Actually, there is a strong tendency of semiconductor crystal to form a covalent bond. So, a hole doesn't tend to exist in a crystal. This can be better understood by the following figure, showing a semiconductor crystal lattice.

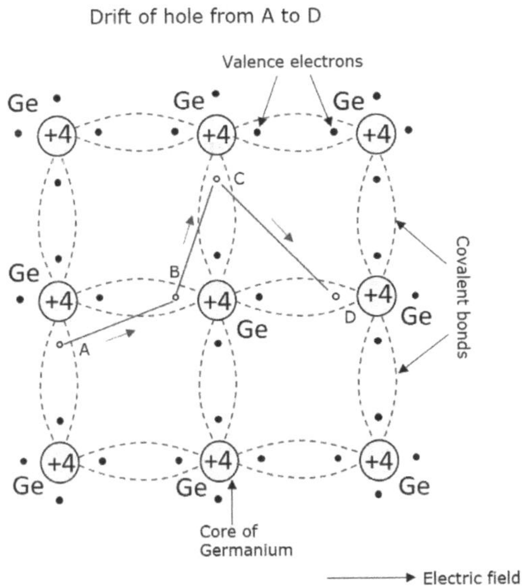

An electron, when gets shifted from a place A, a hole is formed. Due to the tendency for the formation of covalent bond, an electron from B gets shifted to A. Now, again to balance the covalent bond at B, an electron gets shifted from C to B. This continues to build a path. This movement of hole in the absence of an applied field is random. But when electric field is applied, the hole drifts along the applied field, which constitutes the hole current. This is called as hole current but not electron current because, the movement of holes contribute the current flow.

Pentavalent Impurities

Electrons and holes while in random motion, may encounter with each other, to form pairs. This recombination results in the release of heat, which breaks another covalent bond. When the temperature increases, the rate of generation of electrons and holes increase, thus rate of recombination increases, which results in the increase of densities of electrons and holes. As a result, conductivity of semiconductor increases and resistivity decreases, which means the negative temperature coefficient.

Intrinsic Semiconductors

A Semiconductor in its extremely pure form is said to be an intrinsic semiconductor. The properties of this pure semiconductor are as follows –

- The electrons and holes are solely created by thermal excitation.
- The number of free electrons is equal to the number of holes.
- The conduction capability is small at room temperature.

In order to increase the conduction capability of intrinsic semiconductor, it is better to add some impurities. This process of adding impurities is called as **Doping**. Now, this doped intrinsic semiconductor is called as an Extrinsic Semiconductor.

Doping

The process of adding impurities to the semiconductor materials is termed as doping. The impurities added, are generally pentavalent and trivalent impurities.

- The pentavalent impurities are the ones which has five valence electrons in the outer most orbit. Example: Bismuth, Antimony, Arsenic, Phosphorus
- The pentavalent atom is called as a donor atom because it donates one electron to the conduction band of pure semiconductor atom.

Trivalent Impurities

- The trivalent impurities are the ones which has three valence electrons in the outer most orbit. Example: Gallium, Indium, Aluminum, Boron
- The trivalent atom is called as an acceptor atom because it accepts one electron from the semiconductor atom.

Extrinsic Semiconductor

An impure semiconductor, which is formed by doping a pure semiconductor is called as an extrinsic semiconductor. There are two types of extrinsic semiconductors depending upon the type of impurity added. They are N-type extrinsic semiconductor and P-Type extrinsic semiconductor.

N-Type Extrinsic Semiconductor

A small amount of pentavalent impurity is added to a pure semiconductor to result in Ntype extrinsic semiconductor. The added impurity has 5 valence electrons.

For example, if Arsenic atom is added to the germanium atom, four of the valence electrons get attached with the Ge atoms while one electron remains as a free electron. This is as shown in the following figure.

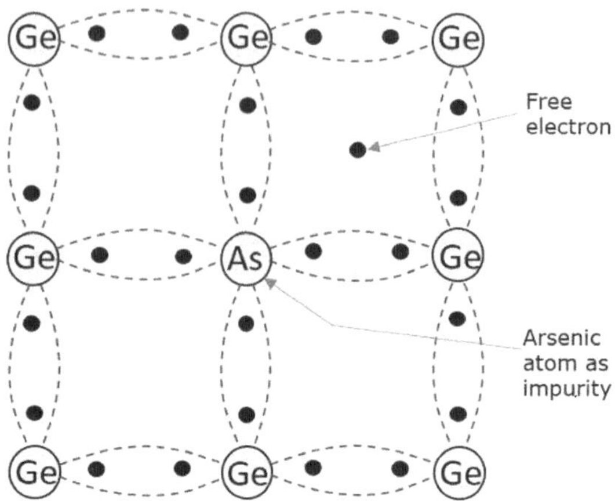

All of these free electrons constitute electron current. Hence, the impurity when added to pure semiconductor, provides electrons for conduction.

In N-type extrinsic semiconductor, as the conduction takes place through electrons, the electrons are majority carriers and the holes are minority carriers.

As there is no addition of positive or negative charges, the electrons are electrically neutral.

When an electric field is applied to an N-type semiconductor, to which a pentavalent impurity is added, the free electrons travel towards positive electrode. This is called as negative or N-type conductivity.

P-Type Extrinsic Semiconductor

A small amount of trivalent impurity is added to a pure semiconductor to result in P-type extrinsic semiconductor. The added impurity has 3 valence electrons. For example, if Boron atom is added to the germanium atom, three of the valence electrons get attached with the Ge atoms, to form three covalent bonds. But, one more electron in germanium remains without forming any bond. As there is no electron in boron remaining to form a covalent bond, the space is treated as a hole. This is as shown in the following figure.

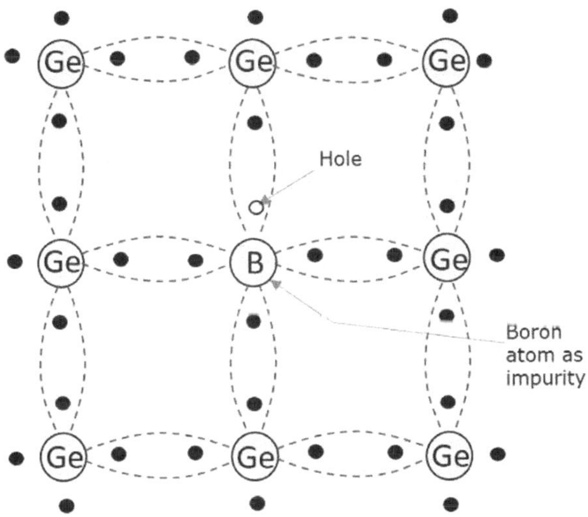

The boron impurity when added in a small amount, provides a number of holes which helps in the conduction. All of these holes constitute hole current.

- In P-type extrinsic semiconductor, as the conduction takes place through holes, the holes are majority carriers while the electrons are minority carriers.
- The impurity added here provides holes which are called as acceptors, because they accept electrons from the germanium atoms.
- As the number of mobile holes remains equal to the number of acceptors, the Ptype semiconductor remains electrically neutral.
- When an electric field is applied to a P-type semiconductor, to which a trivalent impurity is added, the holes travel towards negative electrode, but with a

slow pace than electrons. This is called as P-type conductivity.

- In this P-type conductivity, the valence electrons move from one covalent bond to another, unlike N-type.

Why Silicon is Preferred in Semiconductors?

Among the semiconductor materials like germanium and silicon, the extensively used material for manufacturing various electronic components is Silicon SiSi. Silicon is preferred over germanium for many reasons such as –

- The energy band gap is 0.7ev, whereas it is 0.2ev for germanium.
- The thermal pair generation is smaller.
- The formation of SiO2 layer is easy for silicon, which helps in the manufacture of many components along with integration technology.
- Si is easily found in nature than Ge.
- Noise is less in components made up of Si than in Ge.

Hence, Silicon is used in the manufacture of many electronic components, which are used to make different circuits for various purposes. These components have individual properties and particular uses.

The main electronic components include — Resistors, variable resistors, Capacitors, variable capacitors, Inductors, diodes, Tunnel diodes, Varactor diodes, Transistors, BJTs, UJTs, FETs, MOSFETs, LDR, LED, Solar cells, Thermistor, Varistor, Transformer, switches, relays, etc.

LED

Light-emitting diodes are simply forward-biased p-n junctions, which emit light by spontaneousemission. Spontaneous emission (or electroluminescence) is caused by radiative recombination ofelectron-hole pairs in the depletion region. LED's are temporally and spatially incoherent sourcesthat emit over a broad spectral bandwidth (20-150 nm) and large angular bandwidth.

Advantages of LED

- Simpler fabrication. There are no mirror facets and in some structures no stripedgeometry.
- Cost. The simpler construction of the LED leads to much reduced cost which isalways likely to be maintained.
- Reliability. The LED does not exhibit catastrophic degradation and has proved farless sensitive to gradual degradation than the injection laser. It is also immune toself-pulsation and modal noise problems.
- Generally less temperature dependence. The light output against current characteristic is less affected by temperature than the corresponding characteristic for the injection laser. Furthermore, the LED is not a threshold device and therefore raising the temperature does not increase the threshold current above the operating point and hence halt operation.
- Simpler drive circuitry. This is due to the generally lower drive currents and reduced temperature dependence which makes temperature compensation circuits unnecessary.
- Linearity. Ideally, the LED has a linear light output against current characteristic This can prove advantageous where analog modulation is concerned.

Drawbacks

- Generally lower optical power coupled into a fiber (microwatts);
- Usually lower modulation bandwidth;
- Harmonic distortion.

LED Structures

Types of LED

- Planar LED
- Dome LED
- Surface Emitter LED
- Edge-Emitter LED
- Super luminescent LEDs

Planar LED

The planar LED is the simplest of the structures that are available and is fabricated by either liquid- or vapor- phase epitaxial processes over the whole surface of a GaAs substrate. This involves p-type diffusion into the n- type substrate in order to create the junction. Forward current flow through the junction gives Lambertian spontaneous emission and the device emits light from all surfaces. However, only a limited amount of light escapes the structure due to total internal reflection and therefore the radiance is low.

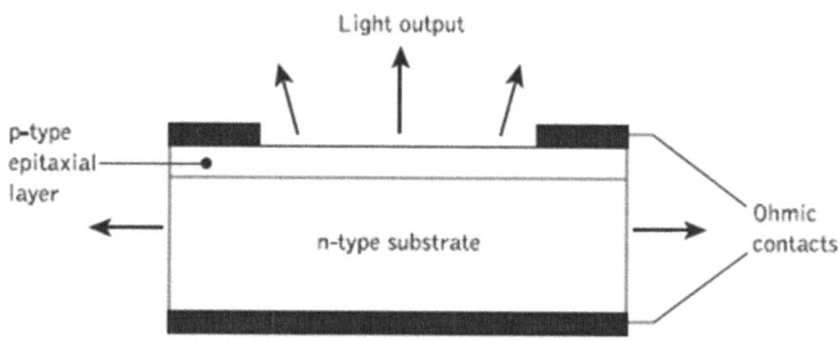

Dome LED

The structure of a typical dome LED is shown in Figure. A hemisphere of n-type GaAs is formed around a diffused p-type region. The diameter of the dome is chosen to maximize the amount of internal emission reaching the surface within the critical angle of the GaAs–air interface. Hence this device has higher external power efficiency than the planar LED. However, the geometry of the structure is such that the dome must be far larger than the active recombination area, which gives a greater effective emission area and thus reduces the radiance.

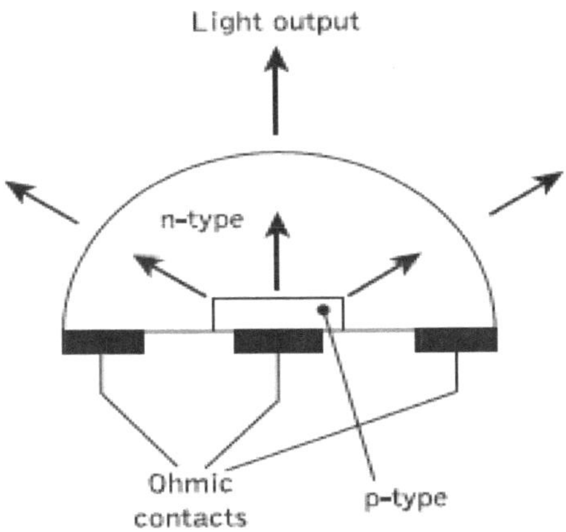

Surface Emitter LED

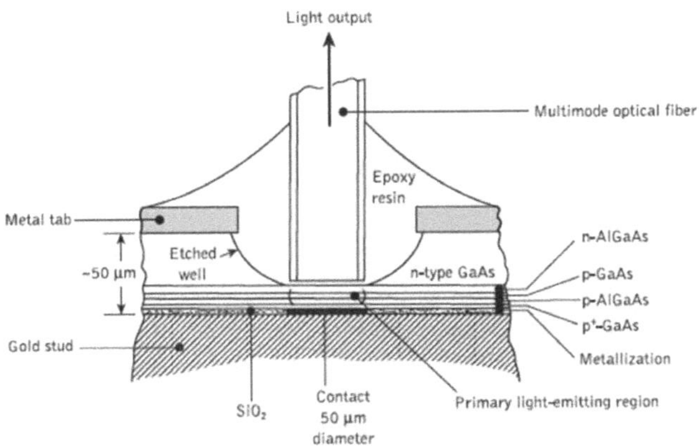

Surface emitter LED (SLED) has been widely employed within optical fiber communications in which A method for obtaining high radiance is to restrict the emission to a small active region within the device. These structures have low thermal impedance in the active region allowing high current densities and giving high- radiance emission into the optical fiber. The structure of a high-radiance etched well DH surface emitter* for the 0.8 to 0.9 μm wavelength band is shown in Figure. The internal absorption in this device is very low due to the larger band-gap-confining layers, and the reflection coefficient at the back crystal face is high giving good forward radiance. The emission from the active layer is essentially isotropic, although the external emission distribution may be considered Lambertian with a beam width of 120° due to refraction from a high to a low refractive index at the GaAs–fiber interface. The power coupled Pc into a multimode step index fiber may be estimated from the relationship:

$$Pc = \pi (1 - r)ARD(NA)^2$$

Where, r is the Fresnel reflection coefficient at the fiber surface, A is the smaller of the fiber core cross-section or the emission area of the source and RD is the radiance of the source.

Edge emitter LED

Edge emitter LED (ELED) has a similar geometry to a conventional contact stripe injection laser

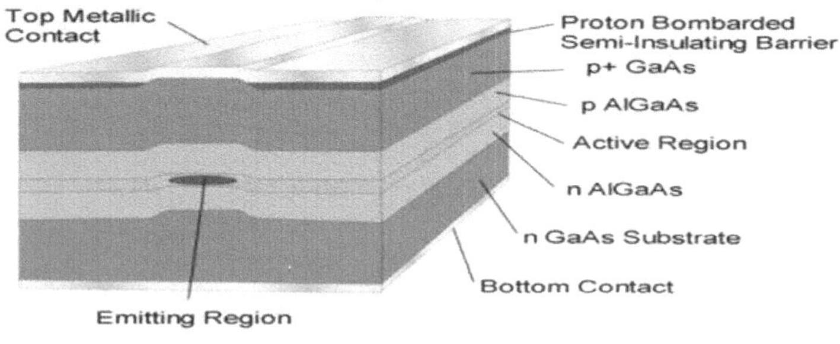

It takes advantage of transparent guiding layers with a very thin active layer (50 to 100 μm) in order that the light produced in the active layer spreads into the transparent guiding layers, reducing self-absorption in the active layer. The consequent wave guiding narrows the beam divergence to a half-power width of around 30° in the plane perpendicular to the junction. However, the lack of wave guiding in the plane of the junction gives a Lambertian output with a half-power width of around 120°.The ELED active layer was heavily doped with Zn to reduce the minority carrier lifetime and thus improve the device modulation bandwidth. In this way a 3 dB modulation bandwidth of 600 MHz was obtained. Very high coupled optical power levels into single-mode fiber in excess of 100 μW have been obtained with In GaAsP ELEDs at drive currents as low as 50 mA.

Super luminescent LED

Another device geometry which is providing significant benefits over both SLEDs and ELEDs for communication applications is the Super luminescent diode or SLD. This device offers advantages of:

- A high output power;
- A directional output beam; and
- A narrow spectral line width.

All of which prove useful for coupling significant optical power levels into optical fiber. The super radiant emission process within the SLD tends to increase the device modulation bandwidth over that of more conventional LEDs.

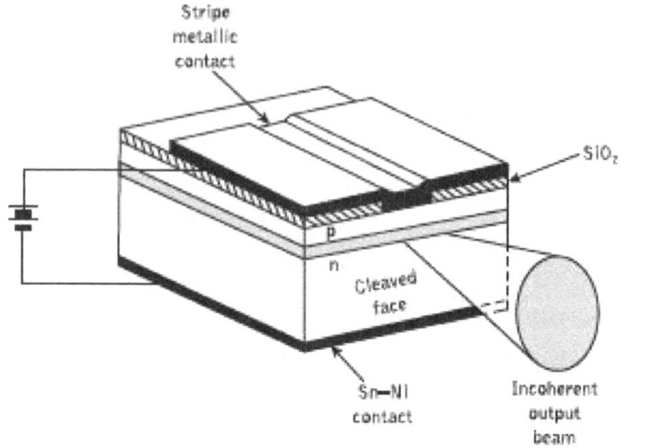

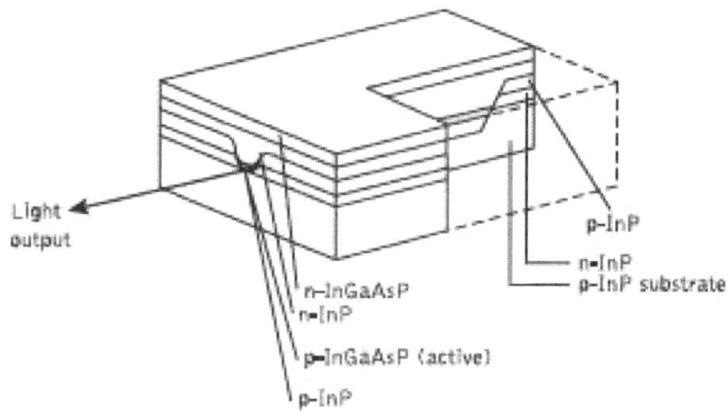

A Super luminescent light emitting diode is, similar to a laser diode, based on an electrically driven pn- junction that, when biased in forward direction becomes optically active and generates amplified spontaneous emission over a wide range of wavelengths. The peak wavelength and the intensity of the SLED depend on the active material composition and on the injection current level. SLEDs are designed to have high single pass amplification for the spontaneous emission generated along the waveguide but, unlike laser diodes, insufficient feedback to achieve lasing action. This is obtained very successfully through the joint action of a tilted waveguide and anti-reflection coated (ARC) facets.

4.14 Light Source Materials

- Light source requirements in optical fiber communication systems light sources play an important role in optical fiber communication systems.
- The basic fiber system consists of transmitter, fiber and receiver.
- The transmitter has a light source that is modulated by a suitable driving circuit depending on the signal to be transmitted.
- The choice of the light source depends on the specific application.
- For high-speed optical fiber communication systems with speeds higher than 1 Gbit/s, the choice of light sources is more critical.
- Several basic requirements should be met by the source.
- The first requirement is that it needs to emit the wavelength corresponding to the loss window of the fused quartz of the most common fiber material, that is, 1. 3um and 1. 5um windows.
- This is very critical because fiber-link roads often operate at a distance of dozens of kilometers without a repeater.
- For the optical power of a given wavelength, a lower fiber loss causes a larger repeater spacing.
- The second requirement is high speed digital modulation.
- At present, the speed of optical fiber communication system has reached 40 Gb/s and 100 Gb/s.
- This requires modulation of the light source at a speed of more than 2. 5Gb/s.

- To meet this requirement, two modulation methods have been developed.
- The first is to directly modulated the light source at the desired speed.
- The second type of modulation is the use of the LiNbO 3 external modulator.
- For the second type, a light source is required to provide a stable power output.
- The next very important feature of this light source is the small spectral width of the light source.
- This will significantly affect the size of the dispersion proportional to the width of the source.
- Splitting in the fiber leads to signal overlap and significantly reduces the bandwidth capacity of the system.
- Although there are many different types of light sources, only LED (LEDs)
- Or laser diodes (LD)
- Because of the requirements mentioned above.
- LED and ld have small size, high power efficiency and many other beneficial features.

Laser diodes (LD)

- Laser is to enlarge the light by the excitation Emission of Radiation.
- The laser is highly monochrome and is conceptually similar to an electronic oscillator.
- The laser consists of an active medium that can provide light amplification and an optical cavity that provides the necessary light feedback.
- The most common laser diode is composed of p-N junction, powered by the injected current.

- It is formed by doping a very thin layer on the surface of the wafer.
- Crystals are doped to produce n-type and a p-
- Type area, one on top of the other, resulting in p-n junction.
- Laser diodes can be used as laser diode modules.
- Some manufacturers offer a wide range of laser diode modules from continuous wave, line generator, modular, NIR and more.
- The semiconductor laser uses a micro-chip of Gallium.
- Arsenide or other foreign semiconductors that produce coherent light in a very small package.
- The energy level difference between the guide band and the valence band electrons provides a mechanism for laser action.
- High power diode laser is the most effective light transmitter.
- They can also be used in laser diode instruments, enabling users to accurately control laser diode current and temperature.
- They can be operated in continuous wave mode by selecting the laser drive current, or they can be modulated by using the modulation function on most drives.
- To accurately stabilize the wavelength, the laser temperature can be fixed.
- The active component is a solid state device that is not completely different from the LED.
- In addition to the critical drive requirements, LD does have some drawbacks.
- Optical properties are generally not equal to other types of lasers.

- In particular, some types of coherence may be lower in length and color.

LEDs (LED)

- LED is p-of forward bias-Where the e-n knot
- H composite through the process of consciously emission led to the generation of light radiation.
- The structure of the LED is similar to the structure of the laser diode, but there is no feedback cavity.
- The emission of the LED is due to spontaneous restructuring, and the output of the LED is very different from the output of the LD (laser diode).
- Led has the advantages of low energy consumption, long life, good robustness, small size and high reliability.
- Unlike the laser diode, there is no threshold, and the output power increases smoothly as the current increases.
- The output power is saturated at high current.
- The total power output from the LEDs can be several milliwatts.
- Since spontaneous radiation is random and appears in all directions, the output of the LED is not directional.
- The angle of the output beam is usually in the 30 ° range perpendicular to the knot, parallel to the approximately 120 ° range of the knot.
- In addition to fiber optic communication, LEDs are also used in many other applications such as aviation lighting, automotive lighting, and traffic signals.
- It isn't just about being on fiber optic cable anymore– it's about maximizing the potential of the platform of manufacturing.

4.15 Quantum Efficiency and LED Power

LED Power and Efficiency

The power generated internally by an LED may be determined by consideration of the excess electrons and holes in the *p*- and *n*-type material respectively. When it is forward biased and carrier injection takes place at the device contacts. The excess density of electrons Δn and holes Δp is equal since the injected carriers are created and recombined in pairs such that charge neutrality is maintained within the structure. In extrinsic materials one carrier type will have a much higher concentration than the other and hence in the *p*-type region, for example, the hole concent ration will be much greater than the electron concentration. Generally, the excess minority carrier density decays exponentially with time t according to the relation:

$$\Delta n = \Delta n(0)\exp(-t/\tau) \quad \text{...... (1)}$$

Where, $\Delta n(0)$is the initial injected excess electron density and τ represents the total carrier recombination lifetime.

When there is a constant current flow into the junction diode, an equilibrium conditions established. In this case, the total rate at which carriers are generated will be the sum of the externally supplied and the thermal generation rates. Hence arate equation for carrier recombination in the LED can be expressed in the form

$$\frac{d(\Delta n)}{dt} = \frac{J}{ed} - \frac{\Delta n}{\tau} \quad (\text{m}^{-3}\,\text{s}^{-1}) \quad \text{.......(2)}$$

The condition for equilibrium is obtained by setting the derivative in Eq. (2) to zero. Hence:

$$\Delta n = \frac{J\tau}{ed} \quad (\text{m}^{-3}) \quad \text{.......(3)}$$

Equation (3) therefore gives the steady-state electron density when a constant current is flowing into the junction region.
It is also apparent from Eq. (2) that in the steady state the total number of carrier recombination per second or the recombination rate *r*t will be:

$$r_t = \frac{J}{ed} \quad (\mathrm{m}^{-3})$$

$$= r_r + r_{nr} \quad (\mathrm{m}^{-3})$$

$$\text{...........(5)}$$

where *r*r is the radiative recombination rate per unit volume and *r*nr is the non-radiative recombination rate per unit volume. Moreover, when the forward-biased current into the device is*i*, then from Eq. (7.4) the total number of recombinations per second *R*t becomes:

$$R_t = \frac{i}{e} \text{............(6)}$$

The LED internal quantum efficiency *η int, which can be defined as the ratio of the radiative recombination rate to the total recombination rate,

$$\eta_{int} = \frac{r_r}{r_t} = \frac{r_r}{r_r + r_{nr}}$$

$$= \frac{R_r}{R_t} \qquad \text{................(8)}$$

Where *R*r is the total number of *radiative* recombination per second. Rearranging Eq. (8) and substituting from Eq.(6)gives:

$$R_r = \eta_{int} \frac{i}{e} \text{.............(9)}$$

Since *R*r is also equivalent to the total number of photons generated per second each photon has an energy equal to *hf* joules, then the optical power generated internally by the LED, *P*int, is:

$$P_{int} = \eta_{int} \frac{i}{e} hf \quad (W)$$

The internally generated power in terms of wave length rather than frequency gives:

$$P_{int} = \eta_{int} \frac{hci}{e\lambda} \quad (W)$$

For the exponential decay of excess carriers depicted by Eq.(1) the radiative minority carrier life time is $\tau r = \Delta n / rr$ and the non-radiative minority carrier lifetime is $\tau nr = \Delta n / rnr$. Therefore, from Eq. (7.7) the internal quantum efficiency is:

$$\eta_{int} = \frac{1}{1 + (r_{nr}/r_r)} = \frac{1}{1 + (\tau_r/\tau_{nr})}$$

Furthermore, the total recombination life time τ can be written as$\tau = \Delta n / r t$gives:

$$\frac{1}{\tau} = \frac{1}{\tau_r} + \frac{1}{\tau_{nr}}$$

Hence,

$$\eta_{int} = \frac{\tau}{\tau_r}$$

4.16 Laser Diodes

What is a Laser Diode?

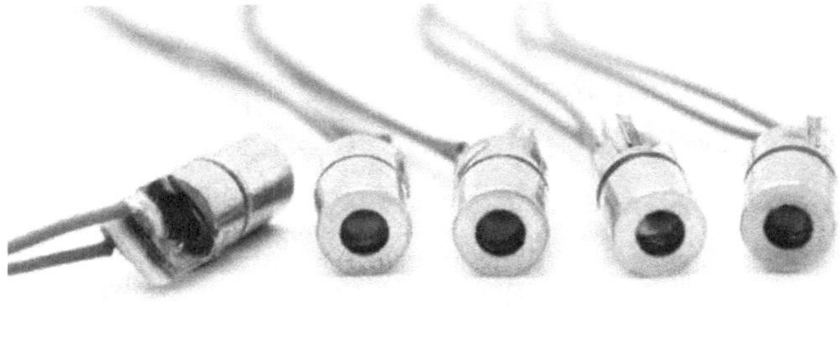

A Laser Diode is a semiconductor device similar to a light-emitting diode (LED). It uses p-n junction to emit coherent light in which all the waves are at the same frequency and phase. This coherent light is produced by the laser diode using a process termed as ―Light Amplification by Stimulated Emission of Radiation‖, which is abbreviated as LASER. And since a p-n junction is used to produce laser light, this device is named as a laser diode. Before we learn more about the working process of a laser diode, let‘s look at how laser light is different from other types of light, and its advantages.

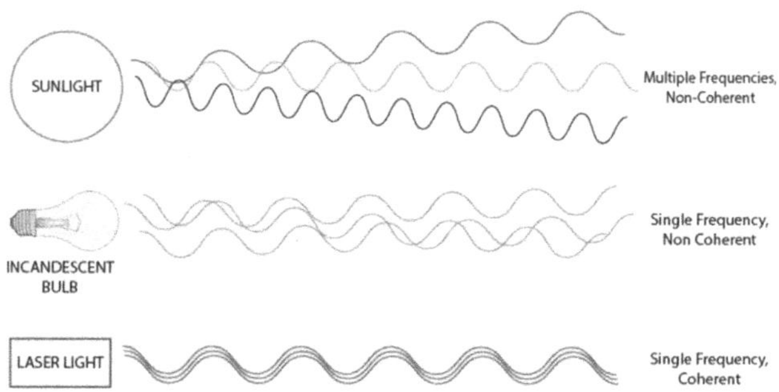

The light from sunlight or from most of the artificial light sources contains waves of multiple wavelengths and they are out of phase with each other. The light waves from monochromatic light sources like incandescent bulb also are not in phase with each other. In contrast to the previous light sources, laser diodes produce a narrow beam of laser light in which all the light waves have similar wavelengths and they travel together with their peaks lined up. This is why laser beams are very bright, and can be focused over a very tiny spot.

Of all the devices that produce laser light, laser diodes or semiconductor lasers are the most efficient and they come in smaller packages. So they are widely used in various devices like laser printers, barcode readers, security systems, Autonomous vehicles (LIDAR), Fiber optic communications etc.

How a Laser Diode Works?

The working of a laser diode takes place in three main steps:

Energy Absorption

The laser diode consists of a p-n junction where holes and electrons exist. (Here, a hole means the absence of an electron). When a certain voltage is applied at the p-n junction, the electrons absorb energy and they transition to a higher energy level. Holes are formed at the original position of the excited electron. The electrons stay in this excited state without recombining with holes for a very small duration of time, termed as ―recombination time‖ or ―upper-state lifetime‖. The recombination time is about a nanosecond for most laser diodes.

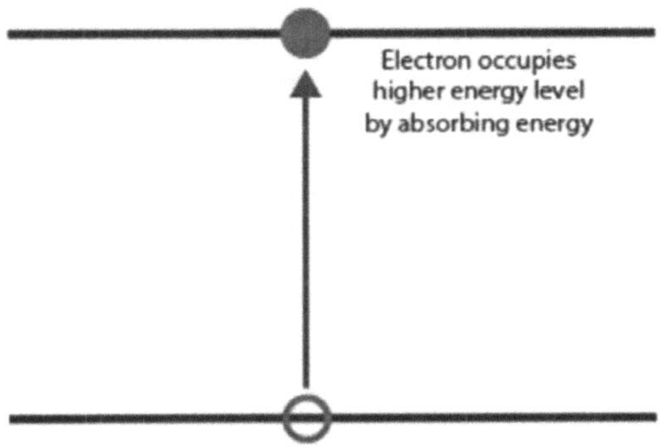

Energy Absorption

Spontaneous Emission

After the upper-state lifetime of excited electrons, they recombine with holes. As the electrons fall from higher energy level to a lower energy level, the difference in energy is converted into photons or electromagnetic radiation. This same process is used to produce light in LEDs. The energy of the emitted photon is given by the difference between the two energy levels.

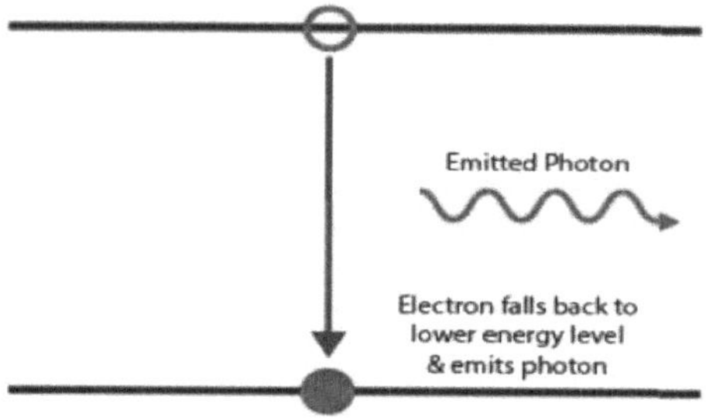

Spontaneous Emission

Stimulated Emission

We need more coherent photons from the laser diode than the ones emitted through the process of spontaneous emission. A partially reflecting mirror is used on either side of the diode so that the photons released from spontaneous emission are trapped in the p-n junction until their concentration reaches a threshold value. These trapped photons stimulate the excited electrons to recombine with holes even before their recombination time. This results in the release of more photons that are in exact phase with the initial photons and so the output gets amplified. Once the photon concentration goes above a threshold, they escape from the partially reflecting mirrors, resulting in a bright monochromatic coherent light.

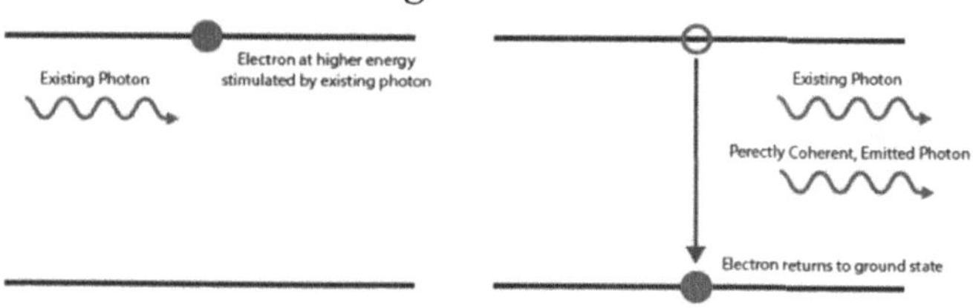

Stimulated Emission

Construction of a Laser Diode

A simple semiconductor laser diode is made up of the following parts in order:

- Metal Contact
- P-type Material
- Active/Intrinsic Region (N-type Material)
- N-type Material
- Metal Contact

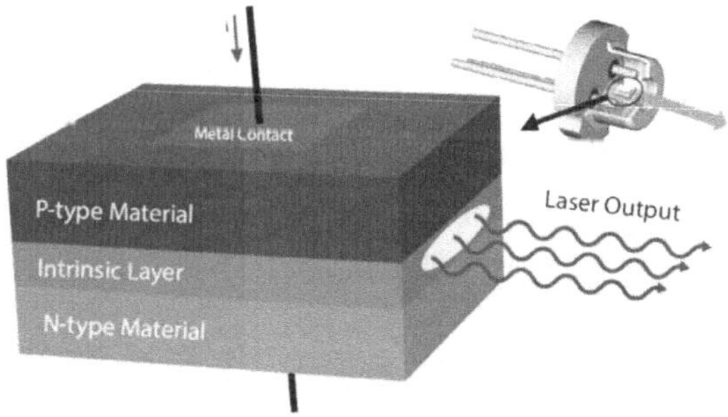

Laser Diode Construction

The input terminals are connected to a metal plates which are sandwiched to the n-type and p-type layers. This type of laser diode is also called as a ―Homojunction Laser Diode‖. The intrinsic region between the p-type and n-type material is used to increase the volume of active region, so that more number of holes and electrons can accumulate at the junction. This allows more number of electrons to recombine with holes at any instant of time, resulting in better output power. The laser light is emitted from the elliptical region. This beam from the laser diode can be further focused using an optical lens. This entire PIN diode (P-type, Intrinsic, N-Type) arrangement is enclosed normally in a metal casing.

Types of Laser Diodes

Double Hetero structure Laser Diode

In this type of laser diodes, an additional confinement layer of a different material is sandwiched between the two p-type and n-type materials. Each of the junction between different materials is called a heterostructure. Because of presence of two heterostructures, this type of laser diode is named as a double heterostructure (DH) laser diode. The advantage of this DH laser diode is that that the active region is confined to a thin layer which gives better optical amplification.

Quantum Well Laser Diode

The quantum well laser diode has a very thin middle layer, which acts as a quantum well. The electrons will be able to use quantum energy levels when transitioning from higher energy level to lower energy level. This gives a better efficiency for this type of laser diode.

Separate Confinement Heterostructure Laser Diode

The thin middle layer in the quantum well laser diode is very small for confining emitted light effectively. To compensate this, in the separate confinement heterostructure laser diode, another two layers are added over the three initial layers. These layers have a lower refractive index and help in confining the emitted light effectively.

Vertical Cavity Surface Emitting Laser Diode (VCSEL)

All the previously discussed laser diodes, the optical cavity is placed perpendicular to the current flow. In vertical cavity surface emitting laser diodes, however, the optical cavity is along the axis of current flow. The partially reflecting mirrors are placed near the ends of optical cavity.

Other Types

- Quantum Cascade Laser Diode
- Interband Cascade Laser Diode
- Distributed Bragg Reflector Laser Diode
- Distributed Feedback Laser Diode
- External Cavity Diode Laser
- Vertical External Cavity Surface Emitting Laser Diode (VCSEL)

Laser Diode P-I Characteristics

The below diagram is a graphical plot between output optical power on y-axis and the current input to the laser diode on x-axis.

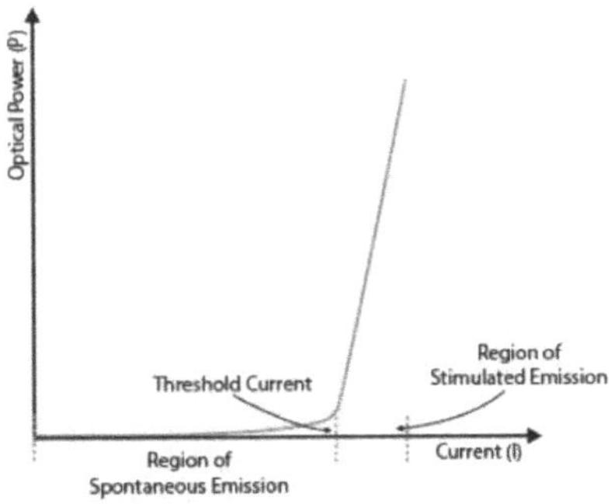

Laser Diaode P – I Graph

As we increase the current flow to the laser diode, the optical power of output light gradually increases up to a certain threshold. Until this point, most of the light emitted is due to spontaneous emission. Above this threshold current, the process of stimulated emission increases. This causes the power of output light to increase a lot even for smaller increases in input current. The output optical power also depends on temperature and it reduces with decrease in temperature.

How to make a Laser Diode Driver Circuit?

Laser diodes require complex drive circuitries that involve feedback loops by measuring output optical power, temperature, voltage and input current. But for controlling a laser diode used in applications where high accuracy is not required, a simple laser diode driver circuit can be constructed using LM317 voltage regulator IC. Below is the diagram.

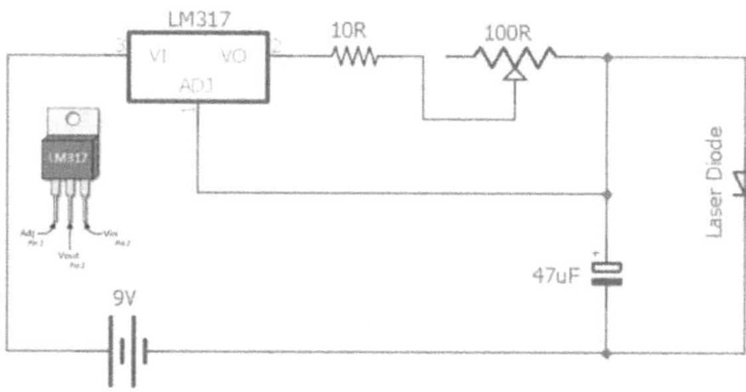

Laser Diode Driver Circuit

The LM317 is configured to function as a constant current source. The output current depends on the value of resistance between Vout and Vadj of LM317 (Pin2 & Pin1). So adjusting the 100R potentiometer will change the output current that flows into the laser diode. The 10R resistor is used to prevent large currents from flowing when the value of 100R potentiometer is at zero. The 47uF capacitor is used to absorb any battery voltage spikes.

Applications of a Laser Diode

Laser Diode modules are used in all major areas of electronics ranging from

Consumer Electronics: CD/DVD players, Laser printers, Fiber Optic Communication, Barcode Readers etc.

Medical Machines: Laser diodes are used in machines used to remove unwanted tissues, eliminating cancer cells, non-invasive and cataract surgeries etc.

Autonomous Vehicles: Laser diode technology is used in making LIDAR systems implemented for autonomous driving

Scientific Instrumentation: Lasers are used in devices used for remote contactless measurements, spectrometry, range finders etc.

Industrial Applications: Laser Diodes are used as a source of high intensity laser beam for precise cutting of materials. They are also used in 3D printing to soften the substrate.

UNIT - V

FIBER OPTICAL RECEIVERS AND DIGITAL TRANSMISSION SYSTEM

Physical principles of photodiodes - PIN photo diode - Avalanche photo diodes - Photodetector noise - SNRDetector response time - Double Hetero structure photodiodes - structure for InGaAS APDs - Temperature effect on avalanche gain - Fundamental receiver operation - Digital signal transmission - Error sources - Front end amplifier - Digital receiver performance - Receiver sensitivity - Optical

5.1 Physical Principles of Photodiodes

A photodiode is a PN-junction diode that consumes light energy to produce an electric current. Sometimes it is also called a photo-detector, a light detector, and photo-sensor. These diodes are particularly designed to work in reverse bias conditions, it means that the P-side of the photodiode is associated with the negative terminal of the battery, and the n-side is connected to the positive terminal of the battery. This diode is very sensitive to light so when light falls on the diode it easily changes light into an electric current. The solar cell is also branded as a large-area photodiode because it converts solar energy into electric energy. Though, solar cell works only in bright light.

What is Photodiode?

A photodiode is one type of light detector, used to convert the light into current or voltage based on the mode of operation of the device. It comprises optical filters, built-in lenses, and also surface areas. These diodes have a slow response time when the surface area of the photodiode increases. Photodiodes are alike to regular semiconductor diodes, but that they may be either visible to let light reach the delicate part of the device. Several diodes intended for use exactly as a photodiode will also use a PIN junction somewhat than the usual PN junction.

Some photodiodes will look like a light-emitting diode. They have two terminals coming from the end. The smaller end of the diode is the cathode terminal, while the longer end of the diode is the anode terminal. See the following schematic diagram for the anode and cathode sides. Under the forward bias condition, the conventional current will flow from the anode to the cathode, following the arrow in the diode symbol. Photocurrent flows in the reverse direction.

Types of Photodiode

Although there are numerous types of photodiode available in the market and they all work on the same basic principles, though some are improved by other effects. The working of different types of photodiodes works in a slightly different way, but the basic operation of these diodes remains the same. The types of photodiodes can be classified based on their construction and functions as follows.

- PN Photodiode
- Schottky Photo Diode
- PIN Photodiode
- Avalanche Photodiode

PN Photodiode

The first developed type of photodiode is the PN type. As compared with other types, its performance is not advanced, but at present, it is used in several applications. The photodetection mainly happens in the depletion region of the diode. This diode is quite small but its sensitivity is not great as compared with others. Please refer to this link to know more about the PN diode.

5.2 PIN Photodiode

At present, the most commonly used photodiode is a PIN type. This diode gathers the light photons more powerfully as compared with standard PN photodiode because the wide intrinsic area between the P and N regions allows for more light to be collected, and in addition to this, it also offers a lower capacitance. Please refer to this link to know more about the PIN diode.

Avalanche Photodiode

This kind of diode is used in low light areas due to its high gain levels. It generates high levels of noise. So this technology is not appropriate for all applications. Please refer to this link to know more about the Avalanche diode.

Schottky Photodiode

The Schottky photodiode uses the Schottky diode, and it includes a small diode junction that means, there is small junction capacitance so, it operates at high speeds. Thus, this kind of photodiode is frequently utilized in high bandwidth (BW) optical communication systems like fiber-optic links. Please refer to this link to know more about the Schottky diode.

Each type of photodiode has its own benefits and drawbacks. The selection of this diode can be done based on the application. The different parameters to be considered while selecting photodiode include noise, wavelength, reverse bias constraints, gain, etc. The performance parameters of photodiode include responsivity, quantum efficiency, transit time or response time.

These diodes are widely used in applications where the detection of the presence of light, color, position, the intensity is required. The main features of these diodes include the following.

- The linearity of the diode is good with respect to incident light
- Noise is low.
- The response is wide spectral
- Rugged mechanically
- Lightweight and compact
- Long life

The required materials to make a photodiode and the range of electromagnetic spectrum wavelength range includes the following

- For silicon material, the electromagnetic spectrum wavelength range will be (190-1100) nm
- For Germanium material, the electromagnetic spectrum wavelength range will be (400-1700) nm
- For Indium gallium arsenide material, the electromagnetic spectrum wavelength range will be (800-2600) nm
- For Lead (II) sulfide material, the electromagnetic spectrum wavelength range will be <1000-3500) nm
- For Mercury, cadmium Telluride material, the electromagnetic spectrum wavelength range will be (400-14000) nm

Because of their better bandgap, Si-based photodiodes produce lower noise than Ge-based photodiodes.

Construction

The photodiode is made using two semiconductors like P-type & N-type. In this design, the formation of P-type material can be done from the diffusion of the P-type substrate which is lightly doped. So, the P+ ions layer can be formed because of the diffusion method. On the substrate of N-type, the N-type epitaxial layer can be grown.

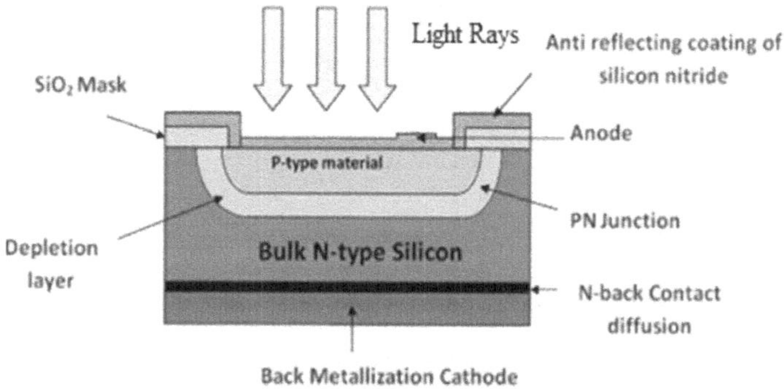

Photodiode Construction

The development of a P+ diffusion layer can be done over the heavily doped N-type epitaxial layer. The contacts are designed with metals to make two terminals like anode and cathode. The front region of the diode can be separated into two types like active & non-active surfaces.

The designing of the non-active surface can be done with silicon dioxide (SiO2). On an active surface, the light rays can strike over it whereas, on a non-active surface, the light rays cannot strike. &the active surface can be covered through the material of anti-reflection so that the energy of light cannot lose and the highest of it can be changed into the current.

Working of Photodiode

The working principle of a photodiode is, when a photon of ample energy strikes the diode, it makes a couple of an electron-hole. This mechanism is also called the inner photoelectric effect. If the absorption arises in the depletion region junction, then the carriers are removed from the junction by the inbuilt electric field of the depletion region.

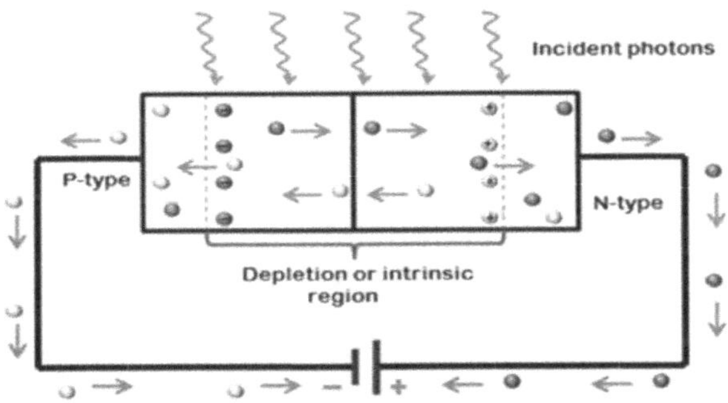

Photodiode Working Principle

Therefore, holes in the region move toward the anode, and electrons move toward the cathode, and a photocurrent will be generated. The entire current through the diode is the sum of the absence of light and the photocurrent. So the absent current must be reduced to maximize the sensitivity of the device.

Modes of Operation

The operating modes of the photodiode include three modes, namely Photovoltaic mode, Photoconductive mode, an avalanche diode mode

Photovoltaic Mode: This mode is also known as zero-bias mode, in which a voltage is produced by the lightened photodiode. It gives a very small dynamic range & non-linear necessity of the voltage formed.

Photoconductive Mode: The photodiode used in this photoconductive mode is more usually reverse biased. The reverse voltage application will increase the depletion layer's width, which in turn decreases the response time & the junction capacitance. This mode is too fast and displays electronic noise

Avalanche Diode Mode: Avalanche diodes operate in a high reverse bias condition, which permits the multiplication of an avalanche breakdown to each photo-produced electron-hole pair. This outcome is an internal gain in the photodiode, which slowly increases the device response.

Why is Photodiode Operated in Reverse Bias?

The photodiode operates in the mode of photoconductive. When the diode is connected in reverse bias, then the depletion layer width can be increased. So this will diminish the capacitance of the junction & the response time. In fact, this biasing will cause quicker response times for the diode. So the relation between photocurrent & illuminance is linearly proportional.

Which is better Photodiode or Phototransistor?

Both the photodiode and phototransistor are used for converting the energy of light to electrical. However, the phototransistor is more responsive as contrasted to the photodiode due to the utilization of the transistor.

The transistor changes the base current which causes due to light absorption & therefore the huge output current can be gained throughout the collector terminal of the transistor. The photodiodes time response is very fast as compared with the phototransistor. So it is applicable where fluctuation in the circuit occurs. For better understating, here we have listed out some points of photodiode vs photoresistor.

Photodiode	Phototransistor
The semiconductor device that converts the energy from light to electrical current is known as a photodiode.	The phototransistor is used to change the energy of light into an electrical current using the transistor.
It generates both the current and voltage	It generates current
The response time is speed	The response time is slow
It is less responsive as compared with a phototransistor	It is responsive and generates a huge o/p current.
This diode works in both the biasing conditions	This diode works in forward biasing only.
It is used in a light meter, solar power plant, etc	It is used to detect the light

Photodiode Circuit

The circuit diagram of the photodiode is shown below. This circuit can be built with a 10k resistor and photodiode. Once the photodiode notices the light, then it allows some flow of current throughout it. The sum of current that supplies through this diode can be directly proportional to the sum of light noticed through the diode.

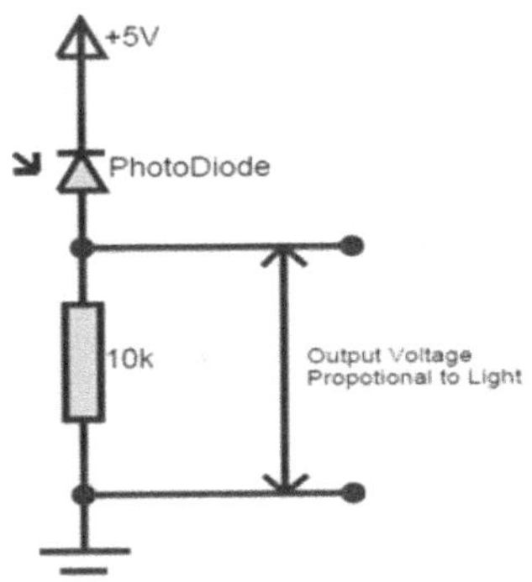

Circuit Diagram

Connecting a Photodiode in an External Circuit

In any application, the photodiode works in reverse bias mode. The anode terminal of the circuit can be connected to the ground whereas the cathode terminal is connected to the power source. Once illuminated through light, then current flows from the cathode terminal to the anode terminal.

Once photodiodes are utilized with exterior circuits, then they are allied to a power source within the circuit. So, the amount of current generated through a photodiode will be extremely small, so this value is not sufficient to make an electronic device.

Once they are connected to an exterior power source, then it delivers more current toward the circuit. In this circuit, the battery is used as a power source to help in increasing the value of current so that external devices give a better performance.

Photodiode Efficiency

The photodiode‘s quantum efficiency can be defined as the division of the absorbed photons which donate to the photocurrent. For these diodes, it is openly associated with the responsivity ‗S‘ with no effect of an avalanche, then the photocurrent can be expressed as

$$I = S\ P = \eta e/h\nu \cdot P$$

Where,

‗η‘ is the quantum efficiency

‗e‘ is the charge of the electron

‗hν‘ is the energy of the photon

Photodiodes‘ quantum efficiency is extremely high. In some cases, it will be above 95% however changes extensively through wavelength. High quantum efficiency requires the control of reflections apart from a high inner efficiency like an anti-reflection coating.

Responsivity

The responsivity of a photodiode is the ratio of the photocurrent which is generated as well as absorbed optical power can be determined within the linear section of the response. In photodiodes, it is normally maximum in a wavelength area wherever the photon energy is fairly higher than the bandgap energy & declining within the bandgap region wherever the absorption reduces.

The photodiode calculation can be done based on the following equation

$R = \eta\,(e/hv)$

Here, in the above equation, ‗h v' is the energy of photon; ‗η' is the efficiency of quantum & ‗e' the charge of elementary. For instance, the quantum efficiency of a photodiode is 90% at an 800 nm wavelength, then the responsivity will be 0.58 A/W.

For photomultipliers & avalanche photodiodes, there is an extra factor for the multiplication of inner current, so that possible values will be above 1 A/W. Generally, the multiplication of current is not included within the quantum efficiency.

PIN Photodiode Vs PN Photodiode

Both the photodiodes like PN & PIN can be attained from a lot of suppliers. A photodiode selection is very important while designing a circuit based on the required performance as well as characteristics.

A PN photodiode doesn't work in a reverse bias & consequently, it is more appropriate for the applications of low light to enhance the performance of noise.

The PIN photodiode that works in reverse bias can introduce a noise current to decrease S/N ratio

For the applications of high dynamic range, the reverse biasing will give good performance

For high BW applications, reverse biasing will provide good performance like the capacitance among the regions of P & N and the storage of charge capacity is small.

Advantages

The advantages of photodiode include the following.

- Less resistance
- Quick and high operation speed
- Long life span.
- Fastest photodetector.
- Spectral response is good.
- Doesn't use high voltage.
- Frequency response is good.
- Solid and low-weight.
- It is extremely responsive to the light.
- Dark current is less.
- High quantum efficiency.
- Less noise.

Disadvantages

The disadvantages of photodiode include the following.

- Temperature stability is poor.
- Change within current is extremely little, therefore may not be enough to drive the circuit.
- The active area is small.
- Usual PN junction photodiode includes a high response time.
- It has less sensitivity.
- It mainly works by depending on the temperature.
- It uses offset voltage.

Applications of Photodiode

- The applications of photodiodes involve similar applications of photodetectors like charge-coupled devices, photoconductors, and photomultiplier tubes.
- These diodes are used in consumer electronics devices like smoke detectors, compact disc players, and televisions and remote controls in VCRs.
- In other consumer devices like clock radios, camera light meters, and street lights, photoconductors are more frequently used rather than photodiodes.
- Photodiodes are frequently used for exact measurement of the intensity of light in science & industry. Generally, they have an enhanced, more linear response than photoconductors.
- Photodiodes are also widely used in numerous medical applications like instruments to analyze samples, detectors for computed tomography, and also used in blood gas monitors.
- These diodes are much faster & more complex than normal PN junction diodes and hence are frequently used for lighting regulation and in optical communications.

V-I Characteristics of Photodiode

A photodiode continually operates in a reverse bias mode. The characteristics of the photodiode are shown clearly in the following figure, that the photocurrent is nearly independent of reverse bias voltage which is applied. For zero luminance, the photocurrent is almost zero excluding for small dark current. It is of the order of nano amperes. As optical power rises the photocurrent also rises linearly. The max photocurrent is incomplete by the power dissipation of the photodiode.

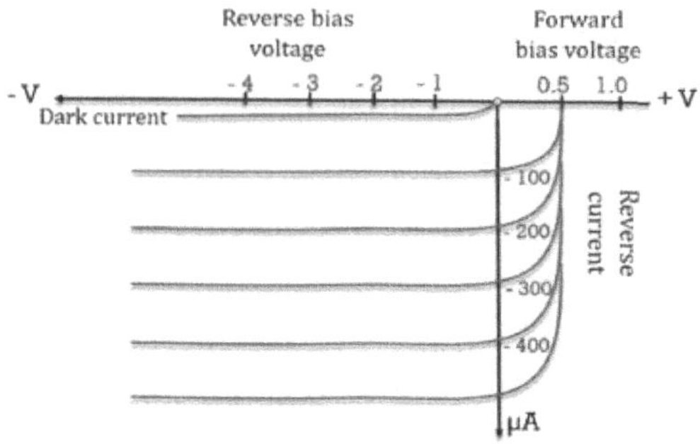

Characteristics

Thus, this is all about the photodiode working principle, characteristics, and applications. Optoelectronic devices like Photodiodes are available in different types which are used in almost all electronic devices. These diodes are used with IR light sources such as neon, laser LED & fluorescent. As compared with other light detection diodes, these diodes are not expensive.

Pin Photo Diode

PIN photodiode is a kind of photo detector, it can convert optical signals into electrical signals.

This technology was invented in the latest of 1950's. There are three regions in this type of diode.

There is a p-region an intrinsic region and an n-region. The p-region and n-region are comparatively heavily doped than the p-region and n-region of usual p-n diodes. The width of the intrinsic region should be larger than the space charge width of a normal pn junction.

The PIN photo diode operates with an applied reverse bias voltage and when the reverse bias is applied, the space charge region must cover the intrinsic region completely. Electron hole pairs are generated in the space charge region by photon absorption. The switching speed of frequency response of photodiode is inversely proportional to the life time.

The switching speed can be enhanced by a small minority carrier lifetime. For the photo detector applications where the speed of response is important, the depletion region width should be made as large as possible for small minority carrier lifetime as a result the switch speed also increases. This can be achieved PIN photodiode as the insertion of intrinsic region the space charge width larger. The diagram of a normal PIN photodiode is given below.

PIN Photodiode

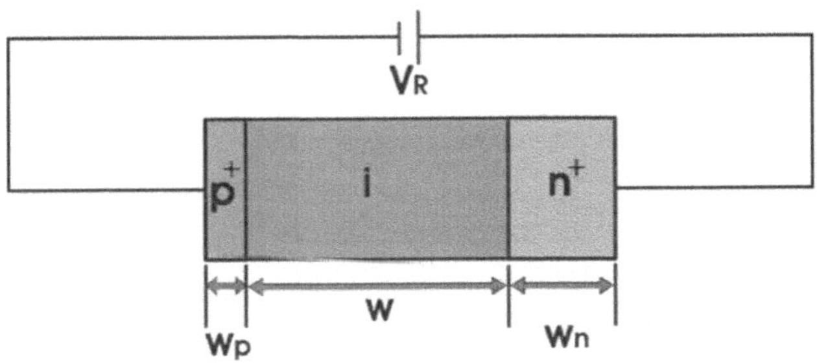

5.3 Avalanche Photodiode

Avalanche photo diode (not to be confused with an avalanche diode) is a kind of photo detector which can convert signals into electrical signals pioneering research work in the development of avalanche diode was done mainly in 1960's.

Avalanche photodiode structural configuration is very similar to the PIN photodiode. A PIN photodiode consists of three regions-

- p-region,
- intrinsic region,
- n-region.

The difference is that reverse bias applied is very large to cause impact ionization. For silicon as the sc material, a diode will need between 100 to 200 volts. Firstly electron- hole pairs are generated by photon absorption in the depletion region. These generate more electron hole pairs through impact ionization. These are swept out of the depletion region quickly, i.e, the transit time is very less.

5.4 Photodetector Noise

Photodetector Noise & S/N

Detection of weak optical signal requires that the photodetector and its following amplification circuitry be optimized for a desired signal-to-noise ratio.

It is the noise current which determines the minimum optical power level that can be detected. This minimum detectable optical power defines the **sensitivity** of photodetector. That is the optical power that generates a photocurrent with the amplitude equal to that of the total noise current (*S/N=1*)

$$\frac{S}{N}=\frac{\text{signal power from photocurrent}}{\text{photodetector noise power}+\text{amplifier noise power}}$$

Signal Calculation

Consider the modulated optical power signal *P(t)* falls on the photodetector with the form of

$$P(t)=P_0[1+ms(t)]$$

Where $s(t)$ is message electrical signal and m is modulation index. Therefore the primary photocurrent is (for pin photodiode $M=1$):

$$i_{ph} = \frac{\eta q}{h\nu} MP(t) = I_p[\text{DC value}] + i_p(t)[\text{AC current}]$$

The root mean square signal current is then

$$\langle i_s^2 \rangle = \langle i_p^2 \rangle M^2 = \sigma_s^2$$

$$\langle i_p^2 \rangle = \sigma_p^2 = \frac{m^2 I_p^2}{2} \quad \text{for sinusoidal signal}$$

5.5 Noise Sources in Photodetectors

The principal noises associated with Photodetectors are :

1. Quantum (Shot) noise: arises from statistical nature of the production and collection of photo-generated electrons upon optical illumination. It has been shown that the statistics follow a Poisson process.
2. Dark current noise: is the current that continues to flow through the bias circuit in the absence of the light. This is the combination of bulk dark current, which is due to thermally generated e and h in the *pn* junction, and the surface dark current, due to surface defects, bias voltage an surface area.
    - In order to calculate the total noise presented in photodetector, we should sum up the root mean square of each noise current by assuming that those are uncorrelated.
    - Total photodetector noise current=quantum noise current +bulk dark current noise + surface current noise

Noise Calculation

Quantum noise current (lower limit on the sensitivity):

$$\langle i_Q^2 \rangle = \sigma_Q^2 = 2qI_p BM^2 F(M)$$

B: Bandwidth, $F(M)$ is the noise figure and generally is

$$F(M) \approx M^x \quad 0 \le x \le 1.0$$

Bulk dark current noise:

$$\langle i_{DB}^2 \rangle = \sigma_{DB}^2 = 2qI_D BM^2 F(M)$$

Surface dark current noise

$$\langle i_{DS}^2 \rangle = \sigma_{DS}^2 = 2qI_L B$$

The total rms photodetector noise current is:

$$\langle i_N^2 \rangle = \sigma_N^2 = \langle i_Q^2 \rangle + \langle i_{DB}^2 \rangle + \langle i_{DS}^2 \rangle$$
$$= 2q(I_P + I_D) BM^2 F(M) + 2qI_L B$$

The thermal noise of amplifier connected to the photodetector is:

$$\langle i_T^2 \rangle = \sigma_T^2 = \frac{4k_B TB}{R_L}$$

S/N Calculation

Having obtained the signal and total noise, the signal-to-noise-ratio can be written as:

$$\frac{S}{N} = \frac{\langle i_P^2 \rangle M^2}{2q(I_P + I_D) BM^2 F(M) + 2qI_L B + 4k_B TB / R_L}$$

Since the noise figure *F(M)* increases with M, there always exists an optimum value of *M* that maximizes the S/N. For sinusoidally modulated signal with *m=1* and *F(M)* = *Mx*

$$M_{opt}^{x+2} = \frac{2qI_L + 4k_B T / R_L}{xq(I_P + I_D)}$$

5.6 SNR detector Response Time

Signal-to-noise ratio (SNR) describes the quality of a measurement. In CCD imaging, SNR refers to the relative magnitude of the signal compared to the uncertainty in that signal on a per-pixel basis. Specifically, it is the ratio of the measured signal to the overall measured noise (frame-to-frame) at that pixel. High SNR is particularly important in applications requiring precise light measurement.

Photons incident on the CCD convert to photoelectrons within the silicon layer. These photoelectrons comprise the signal but also carry a statistical variation of fluctuations in the photon arrival rate at a given point. This phenomenon is known as ―photon noise‖ and follows Poisson statistics. Additionally, inherent CCD noise sources create electrons that are indistinguishable from the photoelectrons. When calculating overall SNR, all noise sources need to be taken into consideration:

Photon noise refers to the inherent natural variation of the incident photon flux. Photoelectrons collected by a CCD exhibit a Poisson distribution and have a square root relationship between signal and noise.

$$(\text{noise} = \sqrt{\text{signal}})$$

Read noise refers to the uncertainty introduced during the process of quantifying the electronic signal on the CCD. The major component of readout noise arises from the on-chip preamplifier.

Dark noise arises from the statistical variation of thermally generated electrons within the silicon layers comprising the CCD. Dark current describes the rate of generation of thermal electrons at a given CCD temperature. Dark noise, which also follows a Poisson relationship, is the square root of the number of thermal electrons generated within a given exposure. Cooling the CCD from room temperature to -25°C will reduce dark current by more than 100 times. In addition, many scientific-grade CCDs employ MPP technology to even further reduce dark current.

Taken together, the SNR for a CCD camera can be calculated from the following equation:

$$\frac{IQEt}{\sqrt{IQEt + Ndt + Nr^2}}$$

where:

I = Photon flux (photons/pixel/second)

QE = Quantum efficiency

t = Integration time (seconds)

Nd = Dark current (electrons/pixel/sec)

Nr = Read noise (electrons)

Under low-light-level conditions, read noise exceeds photon noise and the image data is said to be "read-noise limited". The integration time can be increased until photon noise exceeds both read noise and dark noise. At this point, the image data is said to be "photon limited".

An alternative means of raising the SNR is to use a technique known as binning. Binning is the process of combining charge from adjacent pixels in a CCD during readout into a single "superpixel". Binning neighboring pixels on the CCD array may allow you to reach a photon-limited signal more quickly, albeit at the expense of spatial resolution.

Once you have determined acceptable values for SNR, integration time, and the degree to which you are prepared to bin pixels, the above equation can be solved for the minimum photon flux required. This is, therefore, the lowest light level that can be measured for given experimental conditions and camera specifications.

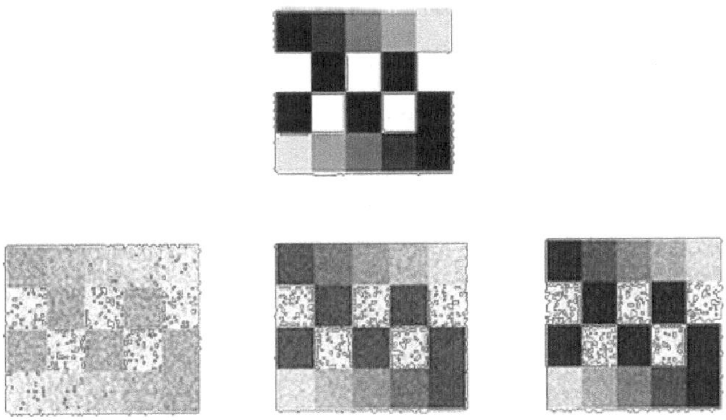

Visual impact of increasing SNR of a typical test pattern

5.7 Double Heterostructure Photodiodes

A double heterostructure is formed when two semiconductor materials are grown into a "sandwich". One material (such as AlGaAs) is used for the outer layers (or cladding), and another of smaller band gap (such as GaAs) is used for the inner layer. In this example, there are two AlGaAs-GaAs junctions (or boundaries), one at each side of the inner layer. There must be two boundaries for the device to be a double heterostructure. If there was only one side of cladding material, the device would be a simple heterostructure.

The double heterostructure is a very useful structure in optoelectronic devices and has interesting electronic properties. If one of the cladding layers is p-doped, the other cladding layer n-doped and the smaller energy gap semiconductor material undoped, a p-i-n structure is formed. When a current is applied to the ends of the pin structure, electrons and holes are injected into the heterostructure. The smaller energy gap material forms energy discontinuities at the boundaries, confining the electrons and holes to the smaller energy gap semiconductor. The electrons and holes recombine in the intrinsic semiconductor emitting photons. If the width of the intrinsic region is reduced to the order of the de Broglie wavelength, the energies in the intrinsic region no longer become continuous but become discrete. (Actually, they are not continuous but the energy levels are very close together so we think of them as being continuous.) In this situation the double heterostructure becomes a quantum well.

5.8 Structure for InGaAa APDS

To improve the performance of structures for InGaAs APDs, various complex device architectures have been devised. One widely used structure is the separate-absorption-and-multiplication (SAM) APD configuration.

Image shows, this structure uses different materials in the absorption and multiplication regions, with each region being optimized for a particular function. Here, light enters the APD through the InP substrate. Since this material has a larger energy band-gap, it allows long-wavelength photons to pass through to the InGaAs absorption region where electron-hole pairs are generated.

Following this is an InP layer that is used for the multiplication region because high electric fields needed for the gain mechanism can exist in InP without tunneling breakdown. This device structure gets its name SAM as a result of the separation of the absorption and multiplication regions.

- Variations on the SAM structure include adding other layers to the device.
- Using a grading layer between the absorption and multiplication regions to increase the response time and bandwidth of the device.
- Adding a charge layer that provides better control of the electric field profile.
- Incorporating a resonant cavity that decouples the optical and electrical path lengths to achieve high quantum efficiencies and wide bandwidths simultaneously.

Another popular design for InGaAs APDs is the superlattice structure. In these devices, the multiplication region is around 250 nm thick and consists of, for example, 13 layers of 9-nm-thick InAIGaAs quantum wells separated by 12-nm-thick InAlAs barrier layers. This structure improves the speed and sensitivity of InGaAs APDs, thereby allowing them to be used for applications such as 10-Gb/s long-distance systems (e.g., SONET OC-192/SDH STM-64 links).

- The transit time of the photocarriers in the depletion region.
- The diffusion time of the photocarriers generated outside the depletion region.
- The RC time constant of the photodiode and its associated circuit.

5.9 Temperature Effect on Avalanche Gain

Large area silicon avalanche photodiodes have been fabricated with maximum avalanche gains exceeding 10,000 and excellent signal to noise ratios. A model of device performance has been developed in which previously developed general expressions are numerically integrated using actual fabrication parameters. The gain, statistical fluctuations in the gain, electronic noise, and total peak broadening have been computed using this model. The results are in good agreement with measurements. The parameter keff was found to be 7.2 X 10-4, allowing a high signal to noise ratio at gains of several thousand.

5.10 Fundamental Receiver Operation

- Digital Signal Transmission
- Error Sources

5.11 Digital Receiver Performance

- Probability of Error
- Receiver Sensitivity
- The Quantum Limit
- Coherent Detection
- Analog Receiver

Optical Receiver Operation

Digital Signal Transmission

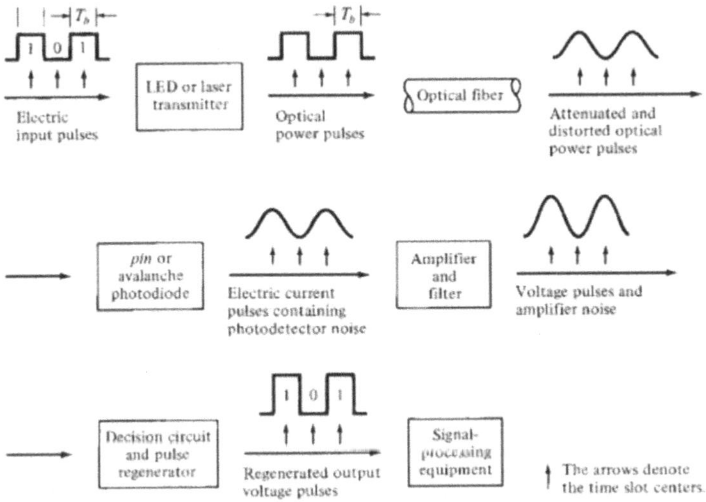

Signal path through an optical data link

- A typical digital fiber transmission link is shown in Figure 4.1. The transmitted signal is a two-level binary data stream consisting of either a ‗0' or a ‗1' in a bit period Tb.
- The simplest technique for sending binary data is amplitude-shift keying, wherein a voltage level is switched between on or off values.
- The resultant signal wave thus consists of a voltage pulse of amplitude V when a binary 1 occurs and a zero-voltage-level space when a binary 0 occurs.
- An electric current i(t) can be used to modulate directly an optical source to produce an optical output power P(t).
- In the optical signal emerging from the transmitter, a ‗1' is represented by a light pulse of duration Tb,whereas a ‗0' is the absence of any light.
- The optical signal that gets coupled from the light source to the fiber becomes attenuated and distorted as

it propagates along the fiber waveguide. Upon reaching the receiver, either a PIN or an APD converts the optical signal back to an electrical format.

- A decision circuit compares the amplified signal in each time slot with a threshold level.
- If the received signal level is greater than the threshold level, a ‗1‘ is said to have been received.
- If the voltage is below the threshold level, a ‗0‘ is assumed to have been received.

5.12 Error Sources

- Errors in the detection mechanism can arise from various noises and disturbances associates with the signal detection system.
- The two most common samples of the spontaneous fluctuations are shot noise and thermal noise.
- Shot noise arises in electronic devices because of the discrete nature of current flow in the device.
- Thermal noise arises from the random motion of electrons in a conductor.
- The random arrival rate of signal photons produces a quantum (or shot) noise at the photodetector. This noise depends on the signal level.
- This noise is of particular importance for PIN receivers that have large optical input levels and for APD receivers.

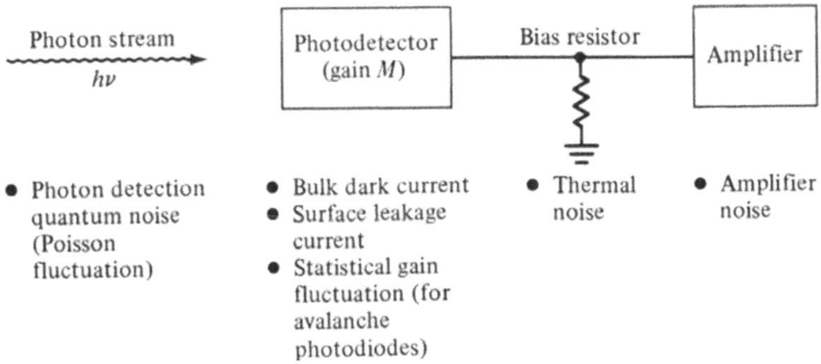

5.13 Noise sources and disturbances in the optical pulse detection mechanism

- When using an APD, an additional shot noise arises from the statistical nature of the multiplication process. This noise level increases with increasing avalanche gain M.
- Thermal noises arising from the detector load resistor and from the amplifier electronics tend to dominate in applications with low SNR when a PIN photodiode is used.
- When an APD is used in low-optical-signallevel applications, the optimum avalanche gain is determined by a design tradeoff between the thermal noise and the gain-dependent quantum noise.
- The primary photocurrent generated by the photodiode is a time-varying Poisson process.
- If the detector is illuminated by an optical signal P(t), then the average number of electron-hole pairs generated in a time t is

$$\overline{N} = \frac{\eta}{h\nu}\int_0^t P(t)dt = \frac{\eta E}{h\nu}$$

where, h is the detector quantum efficiency, hn is the photon energy, and E is the energy received in a time interval .

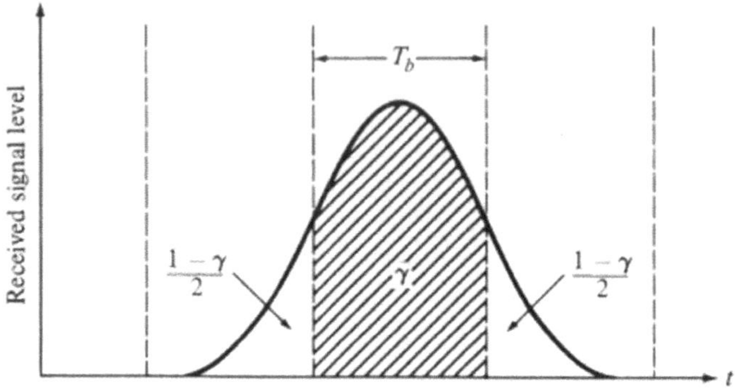

Pulse spreading in an optical signal that leads to ISI

- The actual number of electron-hole pairs n that are generated fluctuates from the average according to the Poisson distribution

$$P_r(n) = \overline{N}^n \frac{e^{-\overline{N}}}{n!}$$

Where, Pr(n) is the probability that n electrons are emitted in an interval t

For a detector with a mean avalanche gain M and an ionization rate ratio k, the excess noise factor F(M) for electron injection is

$$F(M) = kM + \left(2 - \frac{1}{M}\right)(1-k)$$

Or

$$F(M) \simeq M^x$$

Where, the factor x ranges between 0 and 1.0 depending on the photodiode material.

- A further error source is attributed to intersymbol interference (ISI), which results from pulse spreading in the optical fiber.
- The fraction of energy remaining in the appropriate time slot is designated by g, so that 1-g is the fraction of energy that has spread into adjacent time slots.

5.14 Digital Signal Transmission

The shape of a digital signal at different points along an optical link. The transmitted signal is a two-level binary data stream consisting of either a0 or a 1 in a time slot of duration Tb. This time slot is referred to as a bit period. As noted in Chapter 4, electrically there are many ways of sending a given digital message.18-22 One of the simplest techniques for sending binary data is amplitude-shift keying (ASK) or on-off keying (OOK), where a voltage level is switched between two values, which are usually on or off. The resultant signal wave thus consist of a voltage pulse of amplitude V relative to the zero voltage level when a binary 1 occurs and a zero-voltage-level space when a binary 0 occurs. Depending on the coding scheme to be used, a binary 1 may or may not fill the time slot Tb. For simplicity, here we assume that when 1 is sent, a voltage pulse of duration Tb occurs, whereas for 0 the voltage remains at its zero level.

The function of the optical transmitter is to convert the electric signal to an optical receiver operation signal. As shown in, one way of doing this is by directly modulating the light source drive current with the information stream to produce a varying optical output power P(t). Thus, in the optical signal emerging from the LED or laser transmitter, 1 is represented by a pulse of optical power (light) of duration Tp whereas 0 is the absence of any light.

The optical signal that is coupled from the light source to the fiber becomes attenuated and distorted as it propagates along the fiber waveguide. Upon arriving at the end of a fiber, a receiver converts the optical signal back to an electrical format. The basic components of an optical receiver. The first element is either a pin or an avalanche photodiode, which produces an electric current that is proportional to the received power level. Since this electric current typically is very weak, a front-end amplifier boosts it to a level that can be used by the following electronics.

After the electric signal produced by the photodiode is amplified, it passes through a low-pass filter to reduce the noise that is outside of the signal bandwidth. This filter thus defines the receiver bandwidth. In addition, tp minimize the effects of inter-symbol interference (ISI) the filter can reshape the pulses that have become distorted as they travelled through the fiber. This function is called equalization since it equalizes or cancels pulse-spreading effects.

In the final optical receiver module decision circuit samples the signal level at the midpoint of each time slot and compares it with a certain reference voltage known as the threshold level. If the received signal level is greater than the threshold level, 1 is a said to have been received. If the voltage is below the threshold level, 0 is assumed to have been received. To accomplish this but interpretation, the receiver must know where the bit boundaries are. This is done with the assistance of a periodic waveform called a clock, which has a periodicity equal to the bit interval. Thus, this function is called clock recovery or timing recovery.

In some cases, an Optical Receiver Operation preamplifier is placed ahead of the photodiode to boost the optical signal level before photodetection takes place. This is done so that the signal-to-noise ratio degradation caused by thermal noise in the receiver electronics can be suppressed. Compared with other front-end devices, such as avalanche photodiodes or optical heterodyne detectors, an optical preamplifier provides a larger gain factor and a broader bandwidth. However, this process also introduces additional noise to the optical signal. Chapter before addresses optical amplifier and their effects on system performance.

Error Sources

Errors in the detection mechanism can arise from various noises and disturbances associated with the signal detection system. The term noise is used customarily to describe unwanted components of an electric signal that tend to disturb the transmission and processing of the signal in a physical system, and over which we have incomplete control. The noise sources can be either external to the system (switch and power supply transients). Here, we shall be concerned mainly with internal noise, which is present in every communication system and represents a basic limitation on the transmission or detection of signals. This noise is caused by the spontaneous fluctuations of current or voltage in electric circuits. The two most common examples of these spontaneous fluctuations are shot noise and thermal noise19-21. Shot noise arises in electronic devices because of the discrete nature of current flow in the device. Thermal noise arises from the random motion of electrons in a conductor.

The random arrival rate of signal photons produces a quantum (or shot) noise at the photodetector. Since this noise depends on the signal level, it is of particular importance for pin receivers that have large Optical Receiver Operation input levels and for avalanche photodiode receivers. When using an avalanche photodiode, an additional shot noise arises from the statistical nature of the multiplication process. This noise level increases with larger avalanche gain M. Additional photodetector noises come from the dark current and leakage current. These are independent of the photodiode illumination and can generally be made very small in relation to other noise currents by a judicious choice of components.

Thermal noises arising from the detector load resistor and from the amplifier electronics tend to dominate in applications with low signal-to-noise ratio when a pin photodiode is used. When an avalanche photodiode is used in low-optical-signal-level applications, the optimum avalanche gain is determined by a design tradeoff between the thermal noise and the gain-dependent quantum noise.

Probability of Error - Optical Receiver Operation

In practice, there are several ways of measuring the rate of error occurencess in a digital data stream.23-27 Chapter 14 describes some of these methods. A simple approach is to divide the number Ne of errors occurring over a certain time interval t by the number Nt of pulses (ones and zeros) transmitted during this interval. This is called either the error rate or the bit-error rate, which is commonly abbreviated BER. Thus, we have

BER = Ne/Nt = Ne/Bt

Where, B = 1/Tb is the bit rate (the pulse transmission rate). The error rate is expressed by a number, such as 10-9,for example, which states that, on the average, one error occurs for every billion pulses sent. Typical error rates for optical fiber telecommunication systems range from 10-9to 10-12. This error rate depends on the signal-to-noise ratio at the receiver (the ratio of signal power to noise power). The system error rate requirements and the Optical Receiver Operation noise levels thus set a lower limit on the optical signal power level that is required at the photodetector.

To compute the bit-error rate at the receiver, we have to know the probability distribution28-29 of the signal at the equalizer output. Knowing the signal probability distribution at this point is important because it is here that the decision is made as to whether a 0 or a 1 is sent. The shapes of two signal probability distributions.Front end amplifier

5.14 Digital Receiver Performance

The receiver performance is characterized by measuring the BER as a function of the average optical power received. The average optical power corresponding to a BER of 10^{-9} is a measure of receiver sensitivity. The figure below shows the receiver sensitivity measured in various transmission experiments by sending a long sequence of pseudorandom bits (typical sequence length 2^{15} - 1) over a single-mode fiber and then detecting it by using either a p-i-n or an APD receiver. The experiments were performed at the 1.3- or 1.55-μm wavelength, and the bit rate varied from 100 MHz to 10 GHz. The theoretical quantum limit at these two wavelengths is also shown in the figure.

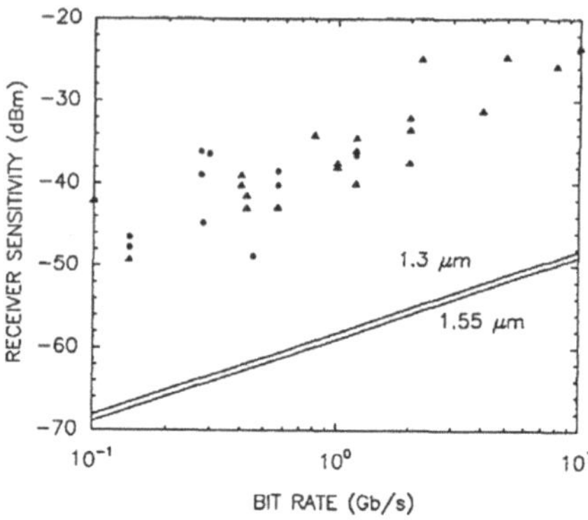

Measured receiver sensitivities versus the bit rate for *p–i–n* (circles) and APD (triangles) receivers in transmission experiments near 1.3- and 1.55-μm wavelengths. The quantum limit of receiver sensitivity is also shown for comparison (solid lines).

A direct comparison shows that the measured receiver sensitivities are worse by 20 dB or more compared with the quantum limit. Most of the degradation is due to the thermal noise that is unavoidable at room temperature and generally dominates the shot noise. Some degradation is due to fiber dispersion, which leads to power penalties; sources of such penalties are discussed in another tutorial.

The dispersion-induced sensitivity degradation depends on both the bit rate B and the fiber length L and increases with BL. This is the reason why the sensitivity degradation from the quantum limit is larger (25-30 dB) for systems operating at high bit rates. The receiver sensitivity at 10 Gb/s is typically worse than -25 dBm. It can be improved by 5-6 dB by using APD receivers. In terms of the number photons/bit, APD receivers require nearly 1000 photons/bit compared with the quantum limit of 10 photons/bit. The receiver performance is generally better for shorter wavelengths in the region near 0.85μm, where silicon APDs can be used; they perform satisfactorily with about 400 photons/bit; an experiment in 1976 achieved a sensitivity of only 187 photons/bit. It is possible to improve the receiver sensitivity by using coding schemes. A sensitivity of 180 photons/bit was realized in a 1.55-µm system experiment after 305 km of transmission at 140 Mb/s.

It is possible to isolate the extent of sensitivity degradation occurring as a result of signal propagation inside the optical fiber. The common procedure is to perform a separate measurement of the receiver sensitivity by connecting the transmitter and receiver directly, without the intermediate fiber. The figure below shows the results of such a measurement for a 1.55-µm field experiment in which the RZ-format signal consisting of a pseudorandom bit stream in the form of solitons (sequence length 2^{23} - 1) was propagated over more than 2000 km of fiber. In the absence of fiber (0-km curve), a BER of 10^{-9} is realized for -29.5 dBm of received power. However, the launched signal is degraded considerably during transmission, resulting in about a 3-dB penalty for a 2040-km fiber link. The power penalty increases rapidly with further propagation. In fact, the increasing curvature of BER curves indicates that the BER of 10^{-9} would be unreachable after a distance of 2600 km. This behavior is typical of most lightwave systems. The eye diagram is related to the use of the RZ format.

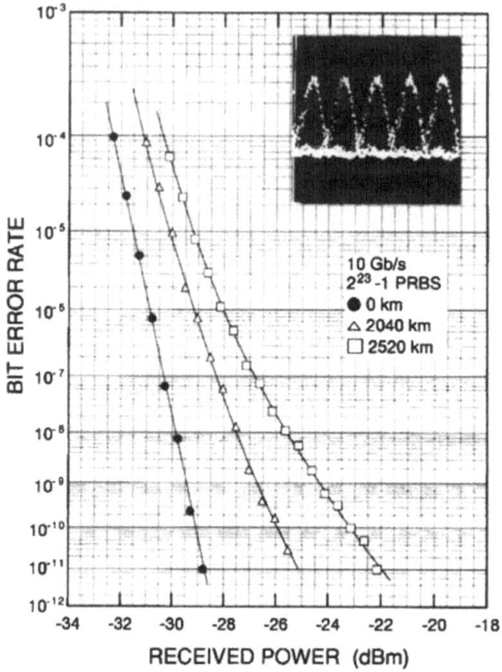

BER curves measured for three fiber-link lengths in a 1.55-μm transmission experiment at 10 Gb/s. Inset shows an example of the eye diagram at the receiver.

The performance of an optical receiver in actual lightwave systems may change with time. Since it is not possible to measure the BER directly for a system in operation, an alternative is needed to monitor system performance. As discussed in another tutorial, the eye diagram is best suited for this purpose; closing of the eye is a measure of degradation in receiver performance and is associated with a corresponding increase in the BER. The eye is wide open in the absence of optical fiber but becomes partially closed when the signal is transmitted through a long fiber link. Close of the eye is due to amplifier noise, fiber dispersion, and various nonlinear effects, all of which lead to considerable distortion of optical pulses as they propagate through the fiber. The continuous monitoring of the eye pattern is common in actual systems as a measure of receiver performance.

The performance of optical receivers operating in the wavelength range 1.3-1.6 μm is severely limited by thermal noise, as seen clearly from the data in the first figure above. The use of APD receivers improves the situation, but to a limited extent only, because of the excess noise factor associated with InGaAs APDs. Most receivers operate away from the quantum limit by 20 dB or more. The effect of thermal noise can be considerably reduced by using coherent-detection techniques in which the receiver signal is mixed coherently with the output of a narrow-linewidth laser. The receiver performance can also be improved by amplifying the optical signal before it is incident on the photodetector.

Ideally, in a digital receiver the decision-circuit output signal voltage vout(t) would always exceed the threshold voltage when a 1 is present and would be less than the threshold when no pulse (a 0) was sent. In actual systems, deviations from the average value of vout(t) are caused by various noises, interference from adjacent pulses, and conditions wherein the light source is not completely extinguished during a zero pulse.

5.15 Receiver Sensitivity

Receivers are integral part of a long distance fiber optic communication system. A receiver includes photodetectors such as Avalanche photodiode, positive-intrinsic-negative semiconductor photodiode etc., demodulators and couplers. Receiver sensitivity is more for an optical receiver when it achieves the same performance with less optical power incident on it. The three important factors which influences receiver sensitivity are bit-error rate (BER), minimum received power and quantum limit of photodetection. Bit-error rate is defined as the probability of incorrect identification of a bit by the decision circuit of the receiver. Minimum received power is a cut-off value below which receiver operation ceases. Use of avalanche photodiode improves receiver sensitivity. But excess noise factor may degrade receiver sensitivity. Quantum limit of photodetection in almost all practical optical receivers is more than 20 dB or exceeds 1000 photons.

5.16 Optical Amplifiers Types

What is an Optical Amplifier?

The transmission loss of the light passing through optical fiber is the very small value of less than 0.2 dB per km with a light wavelength in the 1,550 nm band. However, when the length of the optical fiber is a distance as long as 10 km or 100 km, that transmission loss cannot be ignored. When the light (signal) propagating a long-distance optical fiber becomes extremely weak, it is necessary to amplify the light using an optical amplifier.

An optical amplifier amplifies light as it is without converting the optical signal to an electrical signal, and is an extremely important device that supports the long-distance optical communication networks of today. The major types of optical amplifiers include an EDFA, FRA, and SOA.

5.1 Erbium Doped Fiber Amplifier

Optical amplifier is an optical communication system device. It amplifies an optical signal directly, without converting an optical signal into an electrical signal.

Erbium-doped fiber amplifier (EDFA) is the first successful optical amplifier invented by the UK Southampton University and JP Tohoku University. It is one of the greatest invention in optical communication. Erbium-doped optical fiber is incorporated a small amount of a rare earth element erbium (Er) ion. It is the core of the EDFA. From the late 1980s, the EDFA research has been making a major breakthrough continuously. As WDM technology greatly increases the capacity of optical communication, it becomes the most widely used optical amplifier device in the optical fiber communication.

Principle

EDFA is constituted by a period of erbium-doped fiber (about 10-30m) and pump light source. The stimulated emission of erbium-doped fiber under the action of the pump light source (wavelength 980nm or 1480nm), and the radiation of light varies with the change of the input optical signal, which is equivalent to the input optical signal the amplification. Studies have shown that the erbium-doped fiber amplifiers are typically 15-40dB of gain can be obtained, and the distance relay can be increased on the basis of the original more than 100km. So, why did scientists use erbium-doped fiber element to increase the intensity of light? We know that erbium is a kind of rare earth elements, and rare earth elements has its special structural features. Over the years, people have been using the method which doped rare earth elements in optical devices to improve the performance of optics, so this is not an accidental factor. In addition, why is the pump source wavelength chosen from 980nm or 1480nm? In fact, the pumping light source wavelength could be 520 nm, 650nm, 980nm and 1480nm. But the practice has proved that the 1480nm wavelength pumping light source laser efficiency is the highest, followed by the 980nm wavelength.

Example EDFA Configuration

Irradiating a coupling module with light at 1.48 µm enables the light to be internally stored as energy, and light in the 1.55 μm band causes optical amplification when it propagates, and obtains a gain of 20 to 30 dB.

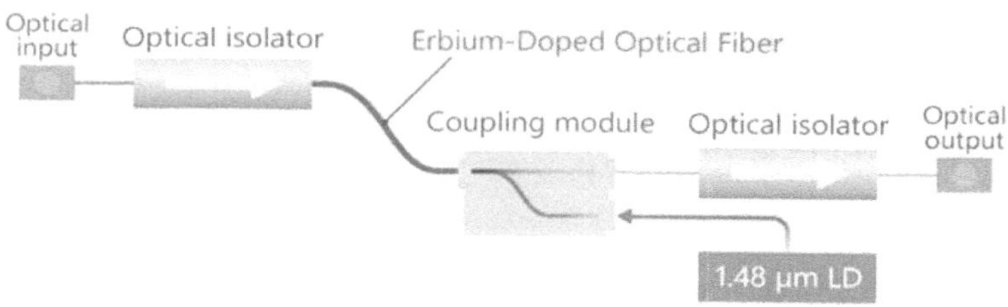

EDFA Configuration

Advantages

The main advantage of EDFA is a high gain, wide bandwidth, high output power, high pumping efficiency, low insertion loss, and not sensitive to the polarization state.

- Its amplifying area happens to coincide with the minimum loss area of single-mode fiber. This reduces the transmission loss of the light signal which can be transmitted relatively far distance.
- It is transparent to digital signal format and data rate.
- Its amplification bandwidth is so wide that dozens or even hundreds of channels can be transmitted in the same fiber.
- It has low noise figure close to the quantum limit, which means that multiple amplifiers can be cascaded.
- Its gain saturation recovery time is long, and has a very small crosstalk between the respective channels.

Applications

When EDFA is used in conventional optical digital communication system applications, we can save a lot of optical repeaters, and the distance relay could also be increased significantly, which is of great significance for the long-haul fiber optic cable trunking systems.

The main applications include:

- It can be used as the light distance amplifier. Traditional electronic fiber optic repeater has many limitations. Such as a digital signal and the analog signal conversion, the repeater should be changed accordingly; repeater changes after the device is changed from a low rate to a high rate; only transmit the same wavelength of the optical signal, and the complex structure, expensive, and so on. Erbium-doped fiber amplifier to overcome these shortcomings, not only do not have to change with the change in the way of the signal, and equipment expansion or for optical wavelength division multiplexing, no need to replace.
- It can be used for the transmitter amplifier and the optical receiver preamplifier. For the rear of the optical transmitter amplifier, the transmit power of the laser is increased from 0dB to +10 db. Optical receiver preamplifier, the sensitivity can also be greatly improved. Therefore, only the line of 1-2 erbium-doped amplifier, the signal transmission distance can be increased to 100-200km. In addition, the erbium-doped fiber amplifier problem to be solved the unique advantages of the erbium-doped fiber amplifier has been recognized by the world, and to be more widely used. However, the erbium-doped fiber amplifier there are also some limitations. For example, in the long-distance communication can not drop channel, each station business contacts is more difficult, not easy to find fault, pumping light source life is not long, as the optical fiber communication technology continues to progress, these problems will be satisfactorily resolved.

FRA (Fiber Raman Amplifier)

An FRA is a type of OFA. It causes stimulated emission based on SRS when strong excitation light enters the optical fiber. The light is then amplified in a wavelength range about 100 nm longer than the excitation light wavelength. It has a wide amplification wavelength region, and can be freely set by the wavelength of the excitation light.

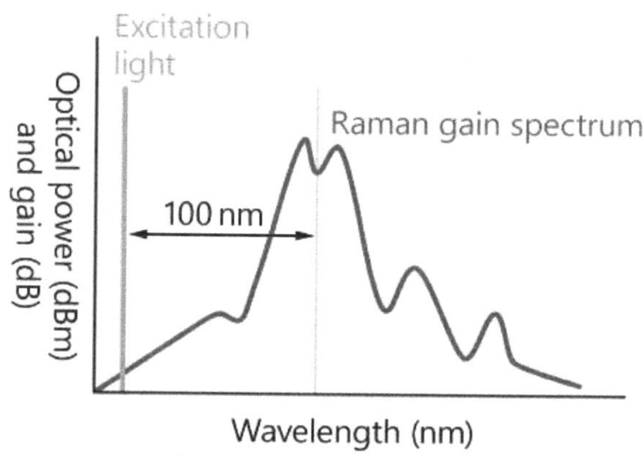

Fiber Raman Amplifier

SOA (Semiconductor Optical Amplifier)

As suggested by the name, an SOA is a semiconductor element. By performing antireflective processing on the cleavage plane of a semiconductor laser and eliminating the resonator structure, light can enter from outside the semiconductor and amplify light via stimulated emission.

An SOA can be made in a compact size and its lower running costs compared to an EDFA mean it more economically efficient. Up until recent years, the input light of an SOA was highly polarization dependent, but research into low polarization dependency has proceeded in recent years. Furthermore, EDFAs are being replaced by SOAs at data centers, and their use is expected to expand in future optical communication.

Example SOA Applications

The SOA is used for embedding in 100G CFP/CFP2 ER4. Light sources for optical communication in the 1.3 μm band is used for 40 km transmission among data centers, and between mobile phone base stations and data centers. However, SOAs are required as pre-amplifiers to counter attenuation in the communication light when performing long-distance transmission. SOAs are embedded in 100 G CFP/CFP2 ER4 transceivers, and are now playing an important role in the market.

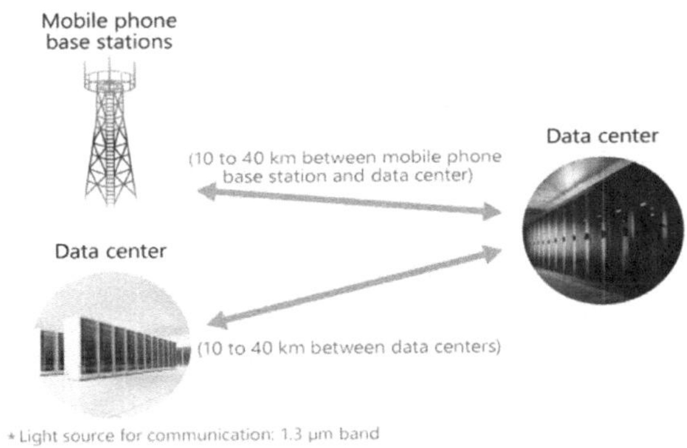

Semiconductor Optical Amplifier

Difference between an OFA and SOA

There are 2 types of optical amplifiers; an OFA (Optical Fiber Amplifier) and SOA (Semiconductor Optical Amplifier).

There are 2 further types of OFAs; an EDFA (Erbium-Doped Fiber Amplifier) and an FRA (Fiber Raman Amplifier).

<table>
<tr><td rowspan="3">Optical amplifiers</td><td rowspan="2">OFA (Optical Fiber Amplifier)</td><td>EDFA (Erbium Doped Fiber Amplifier)</td></tr>
<tr><td>FRA (Fiber Raman Amplifier)</td></tr>
<tr><td colspan="2">SOA (Semiconductor Optical Amplifier)</td></tr>
</table>